The Elements

The Elements

Their Origin, Abundance, and Distribution

P. A. COX

Fellow of New College and
Lecturer in Inorganic Chemistry,
University of Oxford

OXFORD NEW YORK TOKYO
OXFORD UNIVERSITY PRESS
1989

Oxford University Press, Walton Street, Oxford OX2 6DP
Oxford New York Toronto
Delhi Bombay Calcutta Madras Karachi
Petaling Jaya Singapore Hong Kong Tokyo
Nairobi Dar es Salaam Cape Town
Melbourne Auckland
and associated companies in
Berlin Ibadan

Oxford is a trade mark of Oxford University Press

Published in the United States
by Oxford University Press, New York

British Library Cataloguing in Publication Data
Cox, P. A.
The elements: their origin
abundance and distribution.
1. Inorganic compounds. Compounds of main group elements
I. Title
546
ISBN 0–19–855275–0
ISBN 0–19–855298–X Pbk

Library of Congress Cataloging in Publication Data
Cox, P. A.
The elements: their origin, abundance, and distribution/P. A. Cox.
Bibliography: p. Includes indexes.
1. Chemical elements. I. Title.
QD466.C875 1989
546—dc19 88-37342 CIP
ISBN 0–19–855275–0
ISBN 0–19–855298–X (pbk.)

Typeset by Cotswold Typesetting Ltd, Cheltenham
Printed and bound in
Great Britain by Biddles Ltd,
Guildford and King's Lynn

Preface

Fundamental physics has shed a great deal of light on the basic constitution of matter, and even on the origin of fundamental particles—quarks, leptons, photons, and so on—in the Big Bang when the universe as we know it began. These discoveries are fascinating, but they concern forms of matter observable only at very high energies, and somewhat remote from ordinary experience. In practical terms, for most scientists as well as in everyday life, it is the 90 or so naturally occurring chemical elements which are the basic 'building blocks' of ordinary matter. The properties of the elements and their compounds are the subject of innumerable books, most of which—as well as those on the origin of the universe itself—leave unanswered some very fundamental questions. Where do the chemical elements come from? What factors control their very different abundances? How did they end up in the places where we find them today? Such questions should be of interest to chemists and to other physical scientists who study and make use of the properties of the elements. Some partial answers can be found in texts on diverse subjects—inorganic chemistry, astronomy, nuclear physics, and earth sciences—but no one, so far as I am aware, has attempted to write a complete 'natural history' of the elements. This may be because of the wide range of specialist fields involved. In a way, however, it is just this diversity which makes the subject such an interesting one. The story of the elements is inseparable from that of the universe itself, and of the stars and planets which make it up.

The aim of this book is to recount the history of the elements, from their origins in space, through to the minerals and other sources from which we obtain them. There remain many small gaps in this story, but the essential outline is well enough understood. As the account is based on lectures given to first-year university students in chemistry, no specialist knowledge is assumed of fields such as nuclear physics, astronomy, or geology. I have tried to keep even the chemical material fairly descriptive and non-technical in nature, and I hope the book may be suitable for teachers and students in other areas of physical science. Chapter 2 gives an introduction to some of the nuclear and chemical properties of the elements that are essential to the main story. In this and in the later chapters, specialists in the various disciplines covered will no doubt feel that I have simplified the arguments and made light of many of the technical problems which remain. I hope they will excuse this, and will accept that there is some value in a broad introductory approach, uncluttered by detail. For those who wish to pursue some of the topics in more detail, a reading list is provided at the end of each chapter.

I would like to thank my colleagues in Oxford who have helped and encouraged me with this account, and most especially Courtenay Phillips and Bob Denning, whose comments on the manuscript in various drafts were extremely useful. Finally, I must thank my family, Christine, Stephen, Andrew, and Emma, for their own encouragement and understanding.

Oxford P. A. C.
1988

Contents

1
Introduction

I now mean by elements . . . certain primitive and simple, or perfectly unmingled bodies; which not being made of any other bodies, or of one another, are the ingredients of which all those called perfectly mixt bodies are immediately compounded, and into which they are ultimately resolved.
R. Boyle, *The Sceptical Chymist* (1661).

The search for the basic constituents of matter goes back at least to the Greek philosophers of the fifth century BC. Empedocles, who is commemorated in the modern name of his Sicilian birthplace, *Porto Empedocle*, is generally credited with the concept of four basic elements—air, earth, fire, and water—from which everything was supposed to be made in different proportions. This idea was popularized in the writings of Aristotle, and became an accepted part of the medieval world view in Western Europe. It was only during the sixteenth and seventeenth centuries that natural philosophers started to challenge such received ideas, and to look for a more precise view of the nature of matter, based on empirical evidence. The modern idea of a chemical element was formulated in the seventeenth century, and is illustrated by the quotation from Robert Boyle's influential book. The essential distinction between **elements** and **compounds**, in this modern view, is to be made in the laboratory: compounds can be resolved, by suitable chemical procedures, into their constituent elements, whereas elements themselves cannot be separated chemically into simpler substances.

In spite of the great conceptual advances made by Boyle and his contemporaries, it was another century before experimental techniques, developed by such chemists as Black, Priestley, and Lavoisier, had reached the point where elements could be reliably distinguished from compounds. In his *Traité élémentaire de chimie*, published in 1789, Lavoisier correctly identified 30 elements. By the end of the nineteenth century, this list had grown to over 80. Today, it is known that 90 chemical elements occur naturally in larger or smaller amounts on the Earth. Others have been made artificially by nuclear reactions. Most of the elements known on Earth have also been identified as constituents of stars, planets, and other regions of space.

At the same time that many new elements were being identified, Dalton revived the **atomic hypothesis**, another concept which goes back to the Greek philosophers. Dalton proposed that all chemical substances are made of **atoms**, elements containing one type of atom only, and compounds having more than one type present. In the early twentieth century this idea had to be modified slightly: with the discovery of **isotopes**, it was realized that many elements are in fact made of a mixture of atoms with slightly different masses, although very similar chemical properties. Nevertheless, Dalton's essential idea is fundamental to modern thinking on the nature of matter. Each element is characterized by an **atomic number**, and a **relative atomic mass** (often known as its **atomic weight**).

The chemical elements up to number 103 are shown in Table 1.1. This shows their English names, and the internationally agreed symbols which chemists use to represent them. The table also shows the atomic number and the relative atomic mass of each element, and its date of discovery. Several of the elements in the table are listed as being **radioactive**, which means that their atoms are unstable, and change into other elements by emiting high-energy particles. The phenomenon of radioactivity is fundamental to understanding why there is only a finite number of elements. Its causes are explained, along with the physical basis of the atomic number and mass of each element, in Chapter 2. At this point it is worth noting that the last 10 elements in Table 1.1, as well as a few others, decay so rapidly that they are not known naturally on earth, but can only be made artificially. In recent years the list of man-made elements has been extended beyond number 103, and the problems involved in making new elements will also be discussed in Chapter 2.

It is now known that 99 per cent of the universe is made of only two elements, hydrogen and helium. On the other hand, 99 per cent of the mass of the Earth is made up of the eight elements Fe, O, Si, Mg, S, Ni, Ca, and Al. A majority of the elements exist in such small proportions—on Earth and in the universe as a whole—that in most situations they would be regarded as insignificant impurities. Yet nearly all the elements are important to us in different ways. Thirty elements, including some quite rare ones such as selenium, are essential to life. Most elements are used in science and technology, in ways ranging from the large scale (iron, and oxides of silicon and aluminium for construction, for example) to the highly specialized (lithium in medicine and in batteries, platinum as a catalyst, uranium as an energy source). The slow radioactive decay

Table 1.1
The chemical elements

Atomic number	*Name*	*Symbol*	*Relative atomic mass*[a]	*Date of discovery*[b]
1	hydrogen	H	1.008	1766
2	helium	He	4.0026	1895
3	lithium	Li	6.941	1817
4	beryllium	Be	9.01218	1798
5	boron	B	10.81	1808
6	carbon	C	12.011	(ancient)
7	nitrogen	N	14.0067	1772
8	oxygen	O	15.9994	1774
9	fluorine	F	18.99840	1771
10	neon	Ne	20.179	1898
11	sodium	Na	22.9898	1807
12	magnesium	Mg	24.305	1755
13	aluminium	Al	26.98154	1827
14	silicon	Si	28.086	1823
15	phosphorus	P	30.97376	1669
16	sulphur	S	32.06	(ancient)
17	chlorine	Cl	35.453	1774
18	argon	Ar	39.948	1894
19	potassium	K	39.09	1807
20	calcium	Ca	40.08	1808
21	scandium	Sc	44.9559	1879
22	titanium	Ti	47.90	1791
23	vanadium	V	50.9414	1830
24	chromium	Cr	51.996	1797
25	manganese	Mn	24.305	1774
26	iron	Fe	55.847	(ancient)
27	cobalt	Co	58.9332	1774
28	nickel	Ni	58.71	1751
29	copper	Cu	63.546	(ancient)
30	zinc	Zn	65.38	1746
31	gallium	Ga	69.72	1875
32	germanium	Ge	72.59	1886
33	arsenic	As	74.9126	(ancient)
34	selenium	Se	78.96	1817
35	bromine	Br	79.904	1826
36	krypton	Kr	83.80	1898
37	rubidium	Rb	85.4678	1861
38	strontium	Sr	87.62	1790
39	yttrium	Y	88.9059	1794
40	zirconium	Zr	91.22	1789
41	niobium	Nb	92.9064	1801

Table 1.1 (*cont.*)

Atomic number	*Name*	*Symbol*	*Relative atomic mass*[a]	*Date of discovery*[b]
42	molybdenum	Mo	95.94	1778
43	technetium	Tc	98.9062	1937[c]
44	ruthenium	Ru	101.07	1844
45	rhodium	Rh	102.9055	1803
46	palladium	Pd	106.4	1803
47	silver	Ag	107.868	(ancient)
48	cadmium	Cd	112.40	1817
49	indium	In	114.82	1863
50	tin	Sn	118.69	(ancient)
51	antimony	Sb	121.75	(ancient)
52	tellurium	Te	127.60	1782
53	iodine	I	126.9045	1811
54	xenon	Xe	131.30	1898
55	caesium	Cs	132.9054	1860
56	barium	Ba	137.34	1808
57	lanthanum	La	138.9055	1839
58	cerium	Ce	140.12	1803
59	praeseodymium	Pr	140.9077	1879
60	neodymium	Nd	144.24	1885
61	promethium	Pm	145	1946[c,e]
62	samarium	Sm	150.4	1879
63	europium	Eu	151.96	1896
64	gadolinium	Gd	157.25	1880
65	terbium	Tb	158.9254	1843
66	dysprosium	Dy	162.50	1886
67	holmium	Ho	164.9304	1879
68	erbium	Er	167.26	1843
69	thulium	Tm	168.9342	1879
70	ytterbium	Yb	173.04	1878
71	lutetium	Lu	174.97	1907
72	hafnium	Hf	178.49	1923
73	tantalum	Ta	180.9479	1802
74	tungsten	W	183.85	1781
75	rhenium	Re	186.2	1925
76	osmium	Os	190.2	1803
77	iridium	Ir	192.22	1803
78	platinum	Pt	195.09	1735
79	gold	Au	196.9665	(ancient)
80	mercury	Hg	200.59	(ancient)
81	thallium	Tl	204.37	1861
82	lead	Pb	207.2	(ancient)
83	bismuth	Bi	208.9808	1753

Table 1.1 (*cont.*)

Atomic number	*Name*	*Symbol*	*Relative atomic mass*[a]	*Date of discovery*[b]
84	polonium	Po	210	1898[d]
85	astatine	At	210	1940[c]
86	radon	Rn	222	1900[d]
87	francium	Fr	223	1938[c]
88	radium	Ra	226.00254	1898[d]
89	actinium	Ac	227	1899[d]
90	thorium	Th	232.0381	1828[d]
91	protoactinium	Pa	231.0359	1917[d]
92	uranium	U	238.029	1789[d]
93	neptunium	Np	237.0482	1940[c]
94	plutonium	Pu	244	1940[c,e]
95	americium	Am	243	1945[c,e]
96	curium	Cm	247	1944[c,e]
97	berkelium	Bk	247	1950[c]
98	californium	Cf	251	1950[c]
99	einsteinium	Es	254	1952[c,e]
100	fermium	Fm	257	1953[c,e]
101	mendelevium	Md	256	1955[c]
102	nobelium	No	254	1958[c]
103	lawrencium	Lr	257	1961[c]

[a] IUPAC (1983a).
[b] Weeks and Leicester (1968).
[c] Radioactive element first made artificially: occurs naturally (if at all) only in exceedingly small amounts.
[d] Naturally occurring radioactive element.
[e] First discovered in connection with nuclear weapons development: date of publication delayed a year or more after that listed.

of two rare elements (thorium and uranium) and one quite common one (potassium) forms an important source of heat within the Earth, and ultimately fuels many of the geological processes which shape the Earth's surface.

Fundamental to these various roles are not only the chemical properties of each element, but also its abundance and distribution in different regions of the Earth and in space. For example, it is doubtful whether life-forms based on the chemistry of carbon could have developed if this element were as rare as gold. On the other hand, the monetary value of gold depends on just this rarity. It is also notable that gold was known to the ancients, because it is found in nearly pure form in small deposits. However, if the gold in these deposits were distributed uniformly throughout the earth, its concentration would be so low (a few parts per billion) that sophisticated analytical techniques would be required to detect it. Such features of abundance and distribution of the elements are not often considered by chemists, but are clearly important to us all. The main aim of this book is to discuss how they have come about. Yet to pose this problem immediately invites other more fundamental ones:

- Where did the elements come from in the first place?
- How was the Earth formed, with a chemical composition very different from that of the universe as a whole?
- How are the elements distributed within the Earth?
- What geological and chemical processes formed the earth's crust, the surface layer where most of the elements available to us are found?

These questions obviously lead us into fields well outside traditional chemistry: cosmology, astrophysics (for the surprising answer to one of the questions just posed is that most elements are made inside stars), and geology. The reader is not expected to be knowledgeable about any of these fields, as the necessary background will be explained at the appropriate point.

The following section describes the important periodic classification of the elements which will be used later. The remainder of this chapter then looks at what is known 'experimentally' about the abundance and distribution of the elements on Earth and in the universe, ending with a discussion of the types of property required to explain these facts. We shall see at this point that two very different types of behaviour of each element are important: on the one hand, the **chemical properties**, familiar in ordinary laboratory situations, and on the other, the **nuclear properties**, which are necessary to account for the existence of elements and their formation. Chapter 2 contains a brief account of the basis of these two different effects, as a preliminary to the main story of the book, which starts in Chapter 3.

The periodic table of elements

Having found such a large number of 'primitive and simple bodies', it is understandable that chemists should have sought to classify them, and to make some sense out of their very diverse properties. Although there had been earlier attempts, the first successful classification scheme was published in 1869 by Dmitri Mendeleev, and forms the basis of the modern **periodic table of elements**, shown in Fig. 1.1. In this table, elements are listed horizontally in order of their atomic number, forming a series of rows or **periods**. They are arranged to fall vertically in **groups**

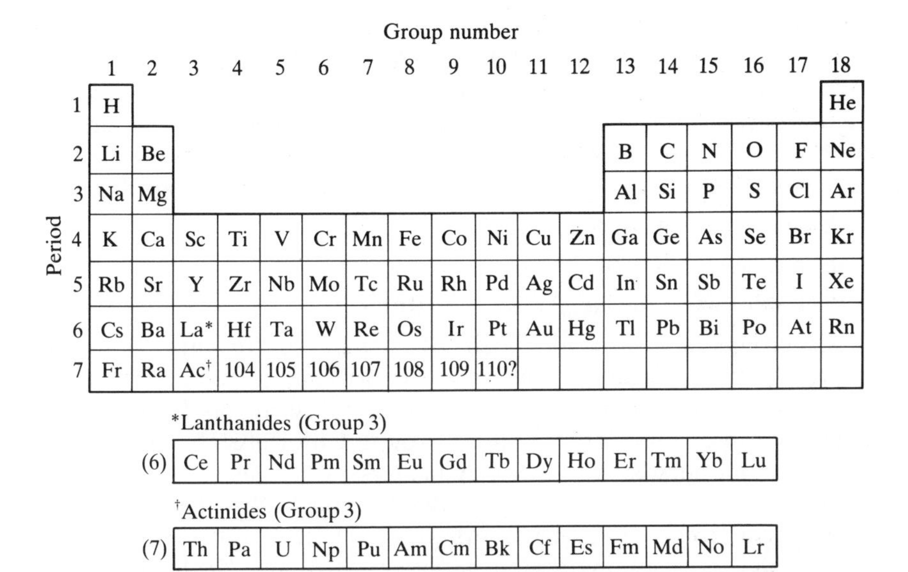

Group number

Period	1	2	3	4	5	6	7	8	9	10	11	12	13	14	15	16	17	18
1	H																	He
2	Li	Be											B	C	N	O	F	Ne
3	Na	Mg											Al	Si	P	S	Cl	Ar
4	K	Ca	Sc	Ti	V	Cr	Mn	Fe	Co	Ni	Cu	Zn	Ga	Ge	As	Se	Br	Kr
5	Rb	Sr	Y	Zr	Nb	Mo	Tc	Ru	Rh	Pd	Ag	Cd	In	Sn	Sb	Te	I	Xe
6	Cs	Ba	La*	Hf	Ta	W	Re	Os	Ir	Pt	Au	Hg	Tl	Pb	Bi	Po	At	Rn
7	Fr	Ra	Ac†	104	105	106	107	108	109	110?								

*Lanthanides (Group 3)

(6)	Ce	Pr	Nd	Pm	Sm	Eu	Gd	Tb	Dy	Ho	Er	Tm	Yb	Lu

†Actinides (Group 3)

(7)	Th	Pa	U	Np	Pu	Am	Cm	Bk	Cf	Es	Fm	Md	No	Lr

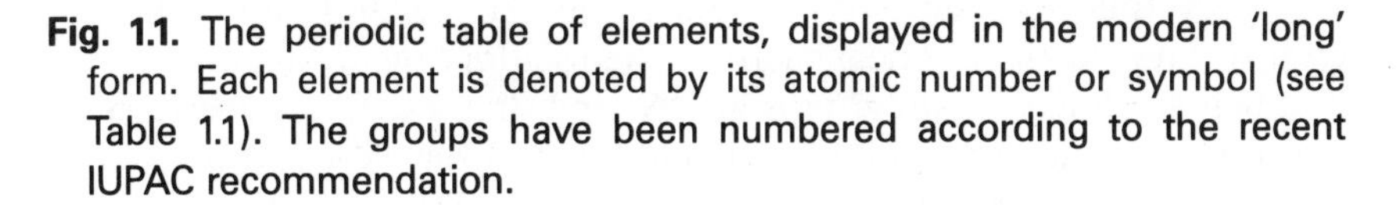

Fig. 1.1. The periodic table of elements, displayed in the modern 'long' form. Each element is denoted by its atomic number or symbol (see Table 1.1). The groups have been numbered according to the recent IUPAC recommendation.

of elements with similar chemical properties, or with regular trends in their properties.

As can be seen in Fig. 1.1 the numbers of elements in the successive periods are:

$$2, \quad 8, \quad 8, \quad 18, \quad 18, \quad 32, \quad (32)$$

the last one being incomplete. These numbers, and the chemical periodicity displayed in the table, can be explained by the quantum theory of atomic structure, which is described briefly in Chapter 2. The fact that periods have different numbers of elements obviously creates problems in constructing the table. The modern 'long' form shown in Fig. 1.1 is to some extent a compromise between the unequal lengths. Ten short groups, containing the **transition elements**, are inserted between the other so-called **main groups**. The fourteen extra elements forming the **lanthanide series** (still referred to often by their older—and inappropriate—name, the rare earths) can be imagined to occupy the same single space as lanthanum in the table. This is justifiable as they are all very similar chemically to lanthanum. In a similar way, fourteen elements in the final period form the **actinide series**, occupying the same place as actinium (although the chemistry of the earlier actinides at least, is not nearly so similar as that of the lanthanides).

Apart from these rather special groups of elements, some other groups in the periodic table are often referred to by names. These are of historical origin, but were intended to express an aspect of their chemical properties. The only group names which we shall use in this book are:

the **alkali metals**: Li, Na, K, Rb, Cs, (Fr);

the **halogens**: F, Cl, Br, I, (At);

the **noble gases**: He, Ne, Ar, Kr, Xe, (Rn).

All groups are also identified by a number. In the traditional nomenclature, the numbers run from I to VIII (or sometimes 0 for the noble gases). The occurrence of the transition elements is accommodated by sub-dividing into 'A' and 'B' groups. Unfortunately (and not for the only time in scientific or other matters), different conventions for these have arisen on the two sides of the Atlantic. In British usage, the 'A' groups run from the left-hand side, up to VIIIA (which is taken to include the three groups beginning with Fe, Co, and Ni, respectively). These are followed (starting with the group Cu, Ag, and Au) by the 'B' groups. The North Americans have come to adopt the different convention that the longer main groups are 'A', and that the 'B' groups refer to the transition elements. Thus, vanadium is in group VA and arsenic in group VB in Britain, but vice versa in the USA. To avoid this confusion (which permeates a great deal of the geochemical literature), the International Union of Pure and Applied Chemistry has recently proposed that all groups should be numbered from 1 to 18. Vanadium is then in group 5, and arsenic in group 15. Although this proposal has been criticized by many chemists on both sides of the Atlantic, we shall adopt it here, and use it in later chapters when elemental distributions are plotted against their position in the periodic table.

The elements on Earth

Nearly all the elements available to us are obtained from minerals in the Earth's **crust**, the layer of rocks forming the top few kilometres of the surface. The most abundant elements in the crust are oxygen and silicon, and for this reason a high proportion of the minerals consist of **silicates**, in which other elements are present in varying proportions. However, this general view hides a picture of very wide diversity. Many elements occur in local concentrations, in 'native' form as metals or alloys, as simple or complex oxides such as carbonates, sulphates, or phosphates, or in other compounds such as sulphides or halides. This diversity reflects both the differences in chemistry between the elements, and the complex processes of fractionation that have accompanied the formation and evolution of the Earth. These processes form the subject matter of Chapters 4 and 5; but the chemical diversity of the crust clearly makes it quite difficult to arrive at an *overall* estimate of the abundance of the different elements. The first serious attempts were made by the American geologists Clarke and Washington, who in the early years of this century published estimates of crustal composition based on the analysis of over 5000 rock specimens. Their results have been refined over the succeeding years, both by using more sophisticated methods of analysing for minor elements, and by taking a better account of the relative proportions of different rock types in the crust. As will be explained in Chapter 5, the composition of the crust also shows quite marked differences in continental and oceanic areas of the Earth.

Figure 1.2 shows a recent estimate of the composition of the continental crust. The average abundance of each element is shown as its mass fraction on a logarithmic scale. This figure shows the very wide range found, extending over more than ten orders of magnitude. Some elements fall off the plot because they are radioactive with no long-lived isotopes: this group contains the heavy elements Po–Ac, Pa, and all the elements beyond U, as well as the two lighter elements Tc and Pm. There

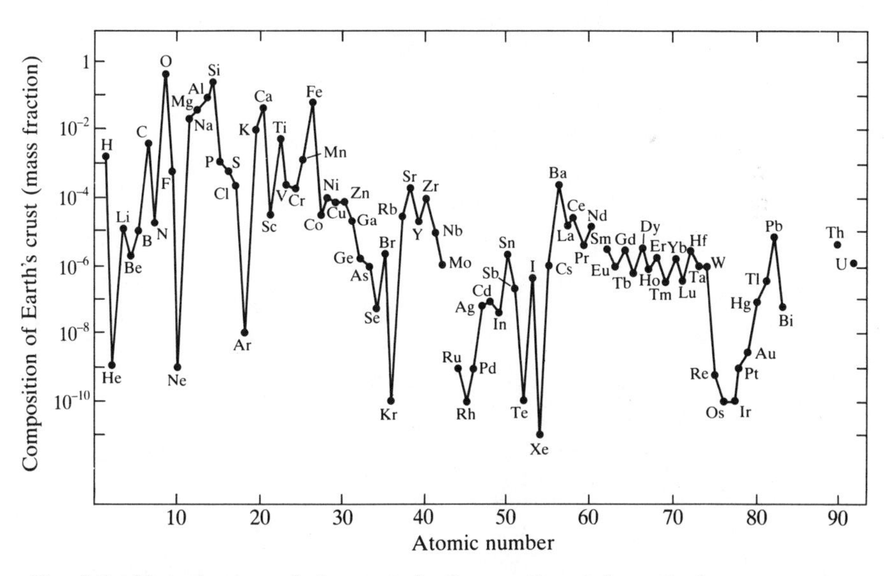

Fig. 1.2. Abundances of elements in the continental crust, shown as mass fraction on a logarithmic scale. (See Appendix A for numerical values and references.)

are also stable elements with crustal abundances of less than one part per billion. Some of these, such as Te and the platinum group elements Ru, Rh, Pd, Os, Ir, and Pt, are probably not as rare on the Earth as a whole (see Fig. 1.4 below), but are concentrated in the central core. For the noble gases He–Xe, on the other hand, the rarity in the Earth's crust almost certainly reflects a very low overall concentration on Earth.

It is now well established that the crust only forms a thin layer, not typical of the structure or composition of the Earth as a whole. The overall density of the Earth, for example (5500 kg/m^3) is higher than can be explained on the basis of silicate minerals throughout. The most detailed information about the internal structure of the Earth is obtained from **seismology**, that is the study of how shock waves, produced by earthquakes or underground nuclear explosions, propagate through the Earth. Figure 1.3 illustrates this structure, and shows that apart from the thin atmosphere and oceans the Earth is composed of three main layers. The principal characteristics of these are also shown in Table 1.2. Beneath the crust is the **mantle**, with a thickness of about 3000 km, and making up around two-thirds of the total mass. Under the mantle is a denser **core**. Seismic studies show that most of the core is liquid, but that there is also a small portion, right at the Earth's centre, which is solid.

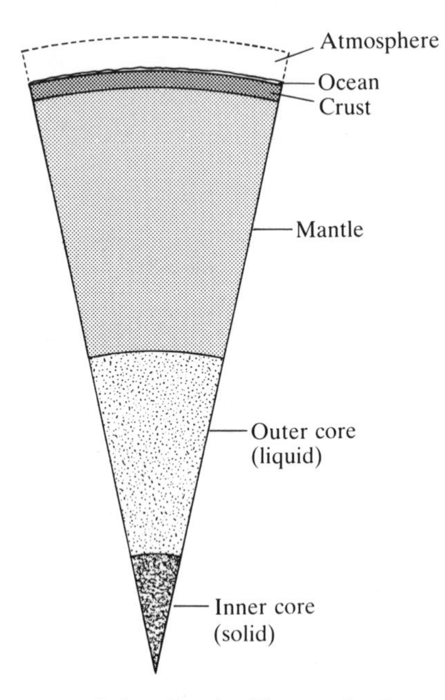

Fig. 1.3. The structure of the Earth. The main features of the different zones are shown in Table 1.2.

Table 1.2
Constitution of the Earth

Component	*Average thickness (km)*	*Average density (10^3 kg/m^3)*	*Fraction of total mass (%)*	*Principal constituents*
Atmosphere	—	—	0.00009	N_2, O_2
Oceans	4	1.03	0.024	H_2O
Crust	17	2.8	0.5	Silicates and other complex oxides
Mantle	2880	4.5	67.2	Mg silicates
Core	3470	11.0	{ 30	Fe, S (liquid)
			2	Fe, Ni (solid)

Although the seismic studies provide some indication of the mechanical properties and the density of the mantle and the core, they give no direct evidence about the chemical composition of these inner parts of the Earth. For this interesting information, therefore, it is necessary to rely on other lines of evidence. Certain rock formations, known as kimberlite pipes, contain diamonds, an allotrope of carbon formed only under very high pressures (around 3.5 GPa, or 35 000 bar). Such high pressures are found at considerable depths inside the Earth, and it is believed that the rocks originate some 100–300 km deep in the mantle. They therefore give some indications about the composition of the upper mantle, which is composed largely of silicate minerals, but with certain important differences from those in the crust. For example, the mantle contains a much higher proportion of magnesium than the crust; on the other hand, the crust is enriched in other elements, including some common ones such as Na, Al, K, and Ca, as well as many minor elements.

No traces of the Earth's core appear at the surface, and models of its composition must come from rather different sources. The density and other physical properties of the core show that it is metallic. Information about the overall composition of the solar system, and especially from meteorites (see the next section) indicates that the abundant element iron is almost certainly the major constituent of the core. This is probably mixed with nearly 10 per cent of nickel, but in order to obtain the correct density there must also be a small percentage of a lighter, non-metallic element. The most likely candidate is sulphur, although other possibilities have been suggested, including oxygen, carbon, and silicon.

Overall estimates of the Earth's composition are still subject to considerable uncertainties, reflecting differences of opinion about the core and the lower regions of the mantle. The data plotted in Fig. 1.4 probably give a good guide, especially for the more abundant elements. As in Fig. 1.2, the mass fraction of each element is plotted on a logarithmic scale. The range of abundances is as wide as for the crust. Only four elements, Fe, O, Si, and Mg, have an abundance of more than 10 per cent, and together they make up 90 per cent of the total mass. Another four elements, S, Ni, Ca, and Al, occur with more than 1 per cent abundance; the remainder together constitute only 1 per cent. About half of the naturally occurring elements, including a few that are essential to life and many others that are used widely by man, have abundances measured in parts per million or less.

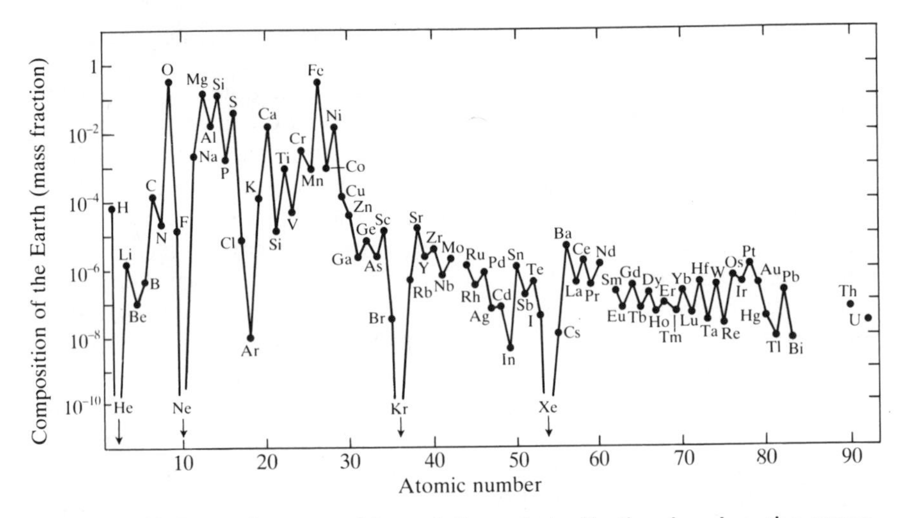

Fig. 1.4. Estimated composition of the whole Earth, showing the mass fraction of each element on a logarithmic scale. (See Appendix A for numerical values and references.)

Although Figs 1.2 and 1.4 show a similar range of abundance, a comparison of the crustal and the whole Earth abundances reveals some obvious differences, reflecting very different distributions of elements between the crust and the inner parts of the Earth. (These are brought out in more detail in Fig. 5.1 on p. 128.) The patterns of elemental distribution on Earth form the basis for an important geochemical classification scheme, illustrated in Fig. 1.5.

Some elements, which are found predominantly in oxide minerals, and to a lesser extent as halides, are much more abundant in the crust than in the inner portions of the Earth. They are known as **lithophiles**, or 'rock-loving' elements.

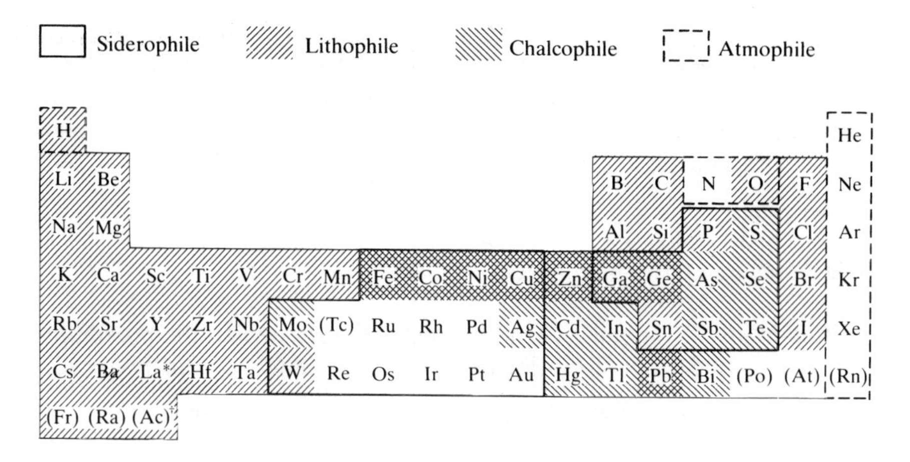

( ) Radioactive element of very low abundance

* Including lanthanides Ce–Lu

† Including actinides Th,U

Fig. 1.5. Geochemical classification of the elements. Lithophiles predominate as oxides and halides in the Earth's crust, siderophiles as metallic alloys in the core. Atmophiles occur largely in volatile form in the atmosphere and oceans. Chalcophiles occur in the crust as sulphides.

The reverse pattern is found for some other elements which tend to be concentrated in the metallic core and are known as **siderophiles** or 'iron-loving' elements.

There are two other, less important classes shown:

atmophiles are those elements (H, N, and the noble gases) which occur largely in volatile form in the atmosphere or oceans;

chalcophiles are elements often found in the crust in combination with the non-metals S, Se, and As, rather than in oxide minerals.

As can be seen from Fig. 1.5 the different classes overlap to some extent. For example, some metals such as Fe, Co, and Ni are predominantly siderophiles, and concentrated in the core. However, they are not uncommon in the crust, and can be found in both oxide and sulphide minerals. There are also a few non-metallic elements such as phosphorus, which in the crust occur exclusively as oxides, but which are probably also present in a reduced chemical form in the core.

The factors underlying the geochemical behaviour of the elements, as well as their abundances, will be explained in later chapters. They reflect differences in chemical affinity which are well known in inorganic chemistry. As we shall see, however, they are also partly a function of the way in which the Earth was formed.

The elements in the solar system

Writing in 1835, the positivist philosopher Comte denied the possibility of ever knowing the composition of the Sun or of other celestial bodies:

> *We understand the possibility of determining their shapes, their distance, their sizes and motions, whereas never by any means will we be able to study their chemical composition, mineralogic structure, and not at all the nature of organic beings living on their surface.*

Yet it is remarkable that we now understand the chemical composition of the Sun and some other parts of the solar system better than that of the Earth.* As explained in the previous section our understanding of the internal composition of the Earth itself comes partly from this source. Until recently, most direct knowledge of the distribution of elements in the solar system came from the Sun—based on spectroscopic observations which ironically enough were started before Comte wrote the lines just quoted, although their significance was not then appreciated—and from meteorites. Recently, this has been supplemented by information obtained by space missions to the Moon and other planets.

The numerous dark lines in the visible spectrum of the Sun were first observed by Wollaston in 1802, and then much more clearly by Fraunhofer a few years later. The interpretation of the Fraunhofer lines, as they are known, came around 1860, largely through the work of Bunsen and Kirchoff. They showed that many of the lines occur at the same wavelengths as ones measured in the laboratory, in the flame spectra of different elements. By 1896 as many as 36 terrestrially known elements had been identified in the solar spectrum. Indeed, one element, helium, was first discovered in this way, before it was known on Earth.

The dark lines in the solar spectrum arise in the relatively cooler,

*Comte is by no means the only philosopher whose predictions have been overtaken by scientific advances. Kant gave strong arguments for the impossibility of our conceiving of a space with non-euclidean geometry—an essential feature of Einstein's general theory of relativity—and even in the twentieth century Wittgenstein predicted that space travel would never be possible.

outer layers of the Sun, when light coming from the inner regions is absorbed by atoms of the different elements present. In principle, the concentration of a given element can be found from the strength of the corresponding absorption, but in practice there are certain difficulties in doing this. The major uncertainty in many cases is the inherent absorption strength, known as the **oscillator strength**, for each line. This is a particular problem with Fe, and for many years the calculated solar abundance of this element was difficult to reconcile with other observations. However, more recent estimates of the oscillator strengths of the Fe lines in the solar spectrum have led to a revision by nearly an order of magnitude in the solar abundance, and the problem has been removed. Certain elements are also relatively difficult to observe, either because of their low abundance, or because (as in the case of the noble gases) their excitation energies are particularly large, so that their lines do not occur in an easily measurable part of the spectrum. Today, however, reasonably good estimates can be made for the abundance of most of the elements, and when these are supplemented by other sources (see below), they form the basis for tabulations of **solar system abundances**. (This term is justified by the fact the Sun itself contains over 99 per cent of the total mass of the solar system.) It is notable that by far the most abundant elements in the Sun are the lightest ones, hydrogen and helium. The atomic ratio H/He is about 10:1, and together they form around 99 per cent of the total matter in the Sun, and thus of the solar system.

The next most important information about the elements in the solar system comes from **meteorites**. These are fragments of rock falling from space, large enough to survive the heating which results from their passage through the atmosphere. They occur in many diverse types, classified for example as 'stones', 'irons', or 'stony irons'. Although their detailed origin is not known, many of these types are thought to come from the break-up of a small planet (or many small planets) which formed in the early stages of the solar system. Thus, they probably resemble the **asteroids**, which form a large number of small planetary bodies, largely found in orbits between those of Mars and Jupiter. It is the composition of the 'iron' types, containing mostly Fe and Ni, which forms the strongest evidence for the nature of the Earth's core (see previous section). The 'stony' meteorites, on the other hand, are largely made up of silicate minerals, and thus resemble the crust and the mantle.

By far the most interesting meteories are those forming a rather rare class known as **carbonaceous chrondrites**. As their name suggests, they contain significant amounts of carbon, which is largely absent in the other types, and is not a very abundant element on Earth. Most meteorites have been subject to periods of considerable heating, which has led to the loss of carbon (as hydrocarbons or as oxides) and other volatile elements. However, the carbonaceous chrondrites have apparently not been so strongly heated, and it is thought that they represent remnants of the original material from which the solar system formed. Some loss of volatile elements has taken place, so that hydrogen and the noble gases are largely missing, and they are somewhat depleted in carbon and oxygen. However, for most elements, the composition of these meteorites is closely similar to that of the Sun. This is illustrated in Fig. 1.6, showing a good correlation between the concentrations of several elements from the two sources. (The elements Li, Be, and B are *more* abundant in meteorites than in the Sun; this is because they are consumed by nuclear reactions in the Sun's interior.) The importance of the correlation is that it enables the pattern of abundances in the Sun to be supplemented, in cases where estimates are impossible or unreliable.

The combination of evidence from the Sun and meteorites allows a fairly good estimate to be made of the solar system abundances of the elements. These are shown in Fig. 1.7 plotted (as the relative numbers of

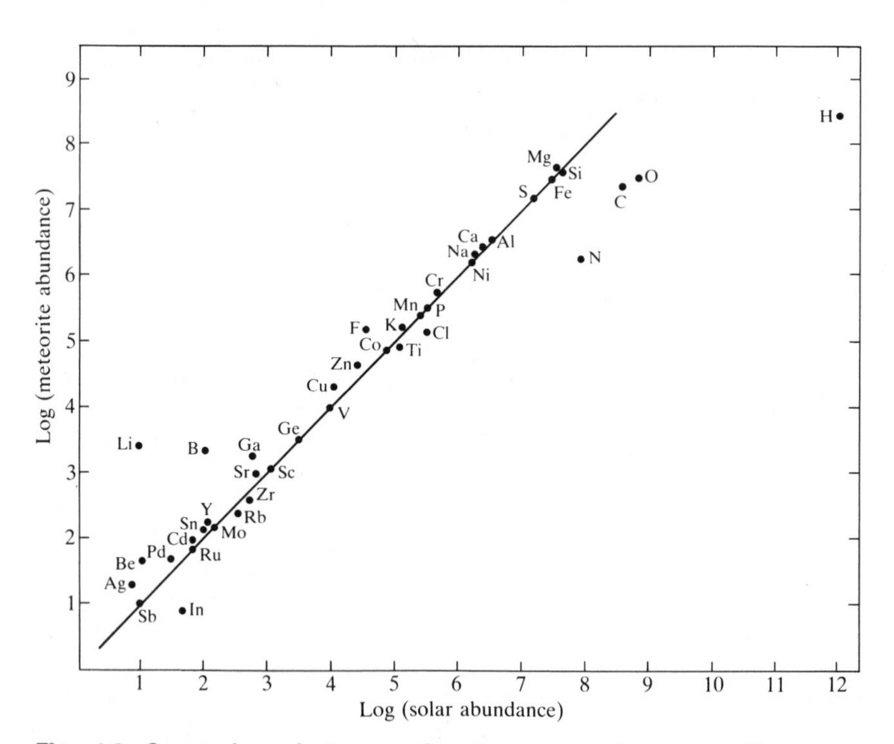

Fig. 1.6. Comparison between abundances of elements with atomic numbers between 1 and 51 found in carbonaceous chondrite meteorites, with those in the Sun. (See Appendix A for data and references.)

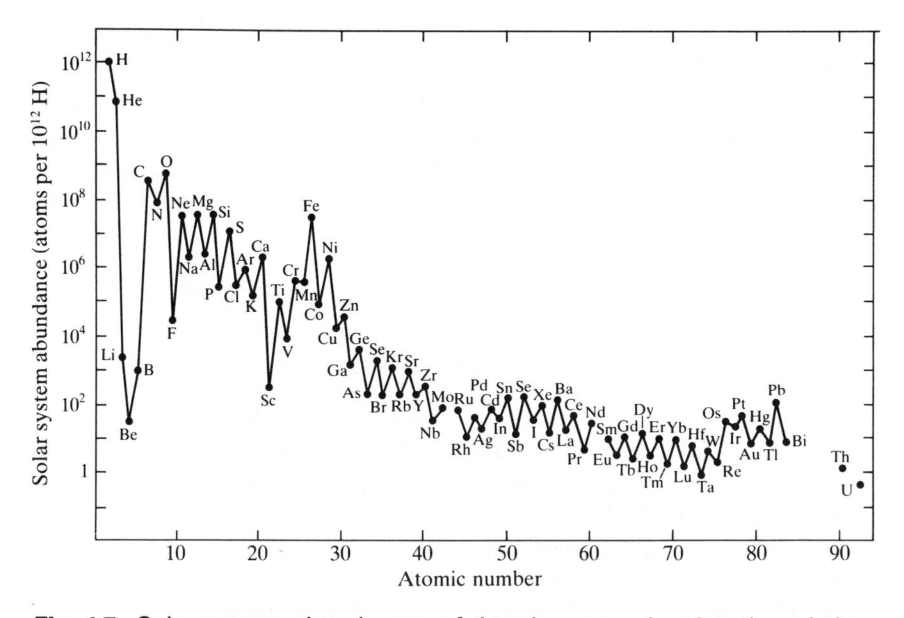

Fig. 1.7. Solar system abundances of the elements, showing the relative number of atoms present on a logarithmic scale, normalized to the value 10^{12} for H. (Based on a combination of solar spectrum and meteorite data: see Appendix A for values and references.)

atoms on a logarithmic scale) as a function of the atomic number. This plot has several notable features. First, as already mentioned, hydrogen and helium make up a very high proportion of the total. There is a general decline in abundance with atomic number, although light elements, such as Li, Be, and B, are all of very low abundance, and there are a number of 'peaks' in the plot, most notably around Fe ($Z = 25$–30). Most striking of all is a pronounced **alternation** of abundances, elements with even atomic numbers being more abundant than those with odd values. These features of the solar system abundances have nothing whatever to do with the chemistry of the elements, but as will be discussed in Chapter 3, are consequences of the nuclear processes by which the elements are synthesized.

Until recently, there was relatively little direct evidence on the compositions of the Moon and of the planets. The densities of most planets can be inferred from their gravitational effects on other planets, and in some cases on their satellites. These densities show that the inner planets, Mercury, Venus, and Mars, together with the Moon, are probably not too dissimilar to the Earth, except that Mercury probably contains a relatively larger iron core, and the Moon rather less core (or none at all). The biggest differences are between this group and the outer planets from Jupiter onwards. These have much lower densities, which show that they must have a very different composition. Some information can be obtained from the spectra of the atmospheres of the 'giant' planets Jupiter and Saturn, showing large amounts of hydrogen, methane, and ammonia. It is thought that hydrogen, indeed, makes up the larger part of these planets, which probably have a composition very similar to the solar-system average. They may have small rocky cores composed of silicates, but a major part of their interiors consists of hydrogen compressed to such a high density that it is metallic.

Much more detailed information about the Moon and the planets has recently become available from manned and unmanned space missions. In addition to rock samples brought back from the Moon, or analysed *in situ* on Mars and Venus, data have also been obtained by remote sensing techniques. Under bombardment from solar radiation and cosmic rays, for example, atoms of different elements emit X-rays of characteristic wavelengths. Orbiting spacecraft, equipped with X-ray spectrometers, can therefore obtain a more complete view of the distribution of elements in a planet's surface than is possible from samples taken from a few, possibly atypical, sites. (This is an interesting extension of an analytical technique widely employed in terrestrial laboratories, known as X-ray fluorescence spectroscopy.) Although the Moon and the inner planets have crusts dominated by the same types of silicate minerals found on Earth, certain differences have been found. For example, many lunar rocks contain higher proportions of the transition elements Ti and Cr than are found on Earth; at the same time, other elements such as Na and K are considerably less abundant. These chemical differences between the planets should be able to provide some of the most important information about the origin of the solar system. They will be discussed in more detail in Chapter 4, where we shall see that in spite of some major advances in understanding, there is still much which is not understood about that interesting event some four-and-a-half billion years ago.

The elements in the universe

Is the solar system typical in its composition or is it some ways unusual, just as the Earth itself has an atypical composition within the solar system? Information similar to that from the sun is available from a very large number of stars, particularly in our own galaxy. By and large, such data support the first possibility: that is, the solar system has a fairly 'typical' composition, similar to that of the universe as a whole—or at least the part of it that we can observe. For this reason, the solar system abundances shown in Fig. 1.7 are sometimes called 'cosmic' abundances. It is important to recognize, however, that there are many differences of detail found in stars. Some of these are quite unusual: for example, exceptional concentrations of heavy elements such as Ba are sometimes seen, and there are even observations of Tc, an element not occurring naturally on Earth because of its radioactive decay. These cases show evidence of the nuclear reactions taking place in stars; they are unusual, in that the surfaces of most stars show little direct sign of such reactions, which normally occur deep in the interior. In addition to the exceptions however, there are some more systematic trends, which appear to give most interesting information about how the elemental composition of the universe has evolved over the course of its lifetime.

Our own solar system forms part of a **spiral galaxy**, containing a total of some 10^{11} stars. Its general shape is illustrated in Fig. 1.8. A high proportion of stars, including the Sun, is concentrated in the plane of the spiral arms. These are known by astronomers as **Population I** stars, and are formed out of material with a very similar composition to the solar system. Out of the plane of the galaxy is a more diffuse **halo** containing many isolated clusters of stars—the so-called 'globular clusters'—classed as **Population II**. Although it is still not understood very well how the spiral structure of a galaxy arises, many observations indicate that the stars in the halo are much older than those in the spiral arms. The Population II stars probably date from the early stages in the formation of the galaxy, more than 10×10^9 years ago. On the other hand, the age of the Sun is about 4.6×10^9 years, and this is typical of Population I stars, although many are much 'younger' still. The crucial observation is that while Population II stars have about the same proportions of hydrogen and helium as the Sun, the amounts of heavier elements appear

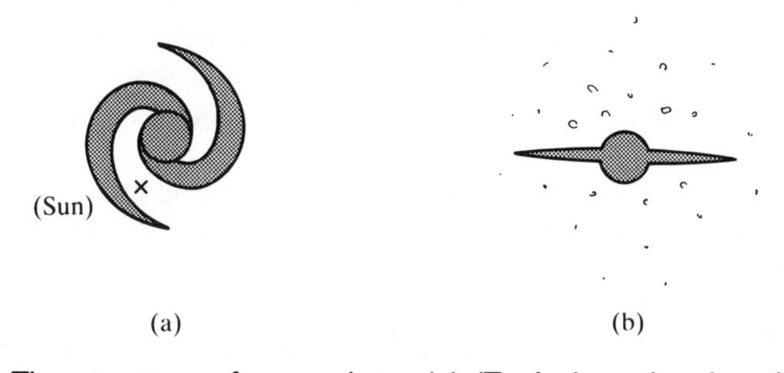

Fig. 1.8. The structure of our galaxy. (a) 'Top' view showing the spiral arms, and the position of the solar system. (b) 'Side' view showing the concentration of stars in the plane of the spiral arms and the diffuse 'halo' of older stars.

to be much less in these older stars, with abundances sometimes as little as 1 per cent of those in the Sun. Since the abundances observed at the surface generally reflect the material from which the star was formed, the differences found between the two populations is clear evidence that the chemical composition of the galaxy has changed in the course of its evolution. This is one line of reasoning which supports the idea, explained in Chapter 3, that the universe was originally made up almost entirely of hydrogen and helium, in proportions that have not changed greatly since, and that the other elements have been synthesized gradually over the course of billions of years.

Although it is the luminous stars that are most easily observed, a significant proportion of the matter contained in the galaxy is dispersed in the form of **inter-stellar clouds.** It has been known for a long time that such clouds contain large amounts of hydrogen; indeed, our knowledge of the structure of the galaxy owes a great deal to the study of an atomic hydrogen emission line at a wave-length of 21 cm by radio astronomers.* More recently, however, many molecules have been detected in space, by their absorption and emission lines in different regions of the spectrum. They include well-known species such as CO and H_2O, as well as many radicals, and some quite large carbon-containing molecules not known in the laboratory. The observation of molecules in space has raised many interesting chemical problems, discussed briefly in Chapter 4. For light atoms, such as H, C, N, and O, they reveal a pattern of elemental abundance quite similar to that in the solar system. Some other common elements, such as Mg, Si, and Fe, appear to be strongly depleted. It is thought that these elements are present in solid form as small dust particles, probably made of silicates and iron metal. It is a pity that the composition of this inter-stellar dust is so difficult to study directly, as it is out of similar material that the Earth and the inner planets were most likely formed. We shall see in Chapter 4 that one can obtain indirect evidence, by assuming that the overall composition of the interstellar medium is similar to that of the solar system, and that the solid dust contains the elements 'missing' from the gas-phase atoms and molecules.

Another source of information on the elements present in distant space comes from **cosmic rays**, which form a background of natural radiation bathing the Earth. The radiation reaching the Earth's surface is made up of secondary particles produced by the primary rays from space in nuclear reactions when they enter the upper atmosphere. The composition of primary cosmic rays themselves was first studied in

*The 21-cm line arises from what is known as a **hyperfine transition**, in which the relative spin directions of the electron and proton change from a parallel orientation to an anti-parallel one.

balloons at high altitudes, and detailed information has been obtained more recently from high-flying aircraft and orbiting satellites. The radiation from space consists largely of atomic nuclei travelling at extremely high velocities. The precise origin of these particles is still uncertain. They may come from supernova explosions of stars, but other high-energy sources, including material orbiting black holes, are possible. Whatever their origin, it is certainly outside the solar system, but probably within our own galaxy. The distribution of elements found in cosmic rays shows broad similarities with that from other sources, but with some important differences. As discussed in Chapter 3 (see Fig. 3.8), the rare elements Li, Be, and B are notably more abundant in cosmic rays, and this gives an important clue as to their origin, which is indeed different from that of most elements.

The factors that control elemental abundance and distribution

Alchemists sought (among other aims) to transmute base metals into gold. With the development of more mechanistic views of the nature of matter this goal was slowly abandoned. Indeed, the **transmutation** of elements one into another cannot, by modern definitions, take place in any chemical reaction. Dalton supposed that atoms were indestructible, and if this is true there is no way in which the elements can have been made; they must have been present right at the origin of the universe, in the same overall proportions as today. However, the discovery of **radioactivity** by Becquerel in 1896, and its subsequent interpretation by Rutherford and Soddy as a form of transmutation process, changed this view of the indestructibility of atoms. A crucial step in understanding was the development of the nuclear model of the atom, by Rutherford in 1911. According to this picture, most of the mass of the atom resides in a small central nucleus, which is positively charged, and is 'orbited' by light, negatively charged electrons rather as in a planetary system. Only the electrons are affected by chemical reactions; the nucleus remains unchanged, and hence preserves the 'chemical identity' of the atom through any chemical change. In this view, then, any transmutation process must change the nuclei of one element into those of another. This can happen (as in some types of radioactive decay) by nuclei splitting into smaller ones, but it is also possible for small nuclei to join together and make larger ones, thus providing a **synthetic** route by which elements can be built up.

We see then, that each element possesses two very different types of property, chemical and nuclear. To understand the origin of the elements, and the overall proportions in which they were made, we need to look at their nuclear reactions. On the other hand, their behaviour on Earth is largely dominated by chemical properties—the relative stabilities of different compounds, for example. We shall see also that the formation of the solar system was dominated by chemical behaviour, and that the way to understand the composition of the Earth and the planets is in terms of **chemical fractionation** of the elements already present.

Because of the importance of both nuclear and chemical properties to our theme, they are discussed in a comparative way in Chapter 2, before taking up the main story. It is worthwhile at this point, however, to get some idea of the contrasting environments in which the two types of process operate, as a result of the very different energy scales involved. Figure 1.9 illustrates this, showing a range of environments covering the temperature span from 1 to 10^{12} K. The right-hand scale shows the average thermal energies of particles at a given temperature. The kinds of species that can exist in the different environments depend roughly on the binding energies of their constituents, relative to this thermal energy.

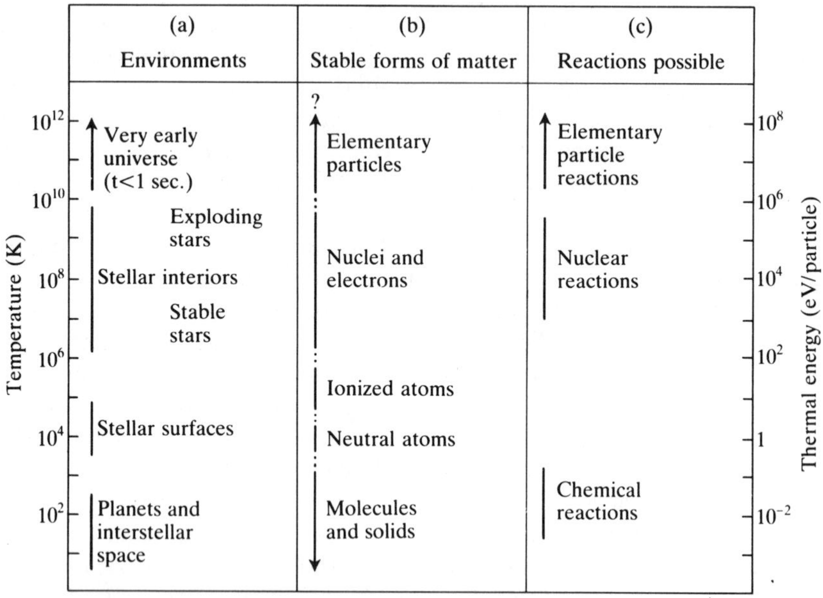

Temperature (K)	(a) Environments	(b) Stable forms of matter	(c) Reactions possible	Thermal energy (eV/particle)
10^{12}	Very early universe (t<1 sec.)	? Elementary particles	Elementary particle reactions	10^{8}
10^{10}	Exploding stars			10^{6}
10^{8}	Stellar interiors; Stable stars	Nuclei and electrons	Nuclear reactions	10^{4}
10^{6}		Ionized atoms		10^{2}
10^{4}	Stellar surfaces	Neutral atoms		1
10^{2}	Planets and interstellar space	Molecules and solids	Chemical reactions	10^{-2}

Fig. 1.9. Energy and temperature scales for chemical and nuclear processes. The scale on the left shows temperature, and that on the right indicates the average thermal energy for the particles present. Column (a) shows typical environments with different temperatures; (b) shows the stable forms of matter present; (c) indicates the types of reaction possible.

Chemical species, in the form of solids and molecules, are restricted to a region below a few thousand degrees. Above this temperature they split first into neutral gas phase atoms, and these in turn split up at higher temperatures into ions and free electrons. The binding energies associated with nuclei are some 10^6 times larger than those of chemical species, and nuclei are stable up to temperatures above 10^9 K. Higher still, however, even nuclei dissociate into elementary particles such as protons and neutrons. Above temperatures of around 10^{12} K even these familiar constituents of matter may be replaced by more exotic particles.

As well as considering the stability of different species, it is also important to look at the possibility of reactions occurring. Except for radioactive decay, which is largely unaffected by temperature or other conditions, most chemical and nuclear processes have 'energy barriers' which must be overcome before a reaction can take place.

Chemical reactions between stable molecules, with barriers of the order of 1 eV (100 kJ/mol, in units more familiar to chemists), proceed at reasonable rates in the temperature range down to 100 K or so. Nuclear reactions have much larger energy barriers, and do not take place at appreciable rates below 10^7 K, unless energy is provided by a non-thermal source—for example, in cosmic rays or man-made particle accelerators. Thus, we can see that chemistry as a phenomenon is restricted to a rather limited range of temperatures—from 10 K, where reactions become very slow, to not too much above 1000 K, where molecules and solids dissociate—whereas nuclear reactions require temperature régimes of 10^7–10^9 K

Various conclusions can be drawn from this discussion. According to the current theory, for which there is now almost overwhelming evidence, the universe began around 16 billion years ago in a state of extraordinarily high temperature. No nuclei, let alone atoms and molecules, could have been present under such conditions. Thus, the chemical elements must indeed have originated somewhere in the universe, and were not present already 'in the beginning'. This origin must have involved nuclear reactions, at temperatures and energies far higher than those of the 'chemical régime'. Chemical processes, therefore, can have played no part whatever at this stage. The chemical differentiation accompanying the formation of the Earth and the planets must have occurred at much lower temperatures. Under such conditions, the nuclear reactions involved in the original formation of the elements must have ceased (although the slow radioactive decay of some elements continued and does so today).

The separation between the régimes where chemical and nuclear processes operate determines the sequence of the chapters which follow. Chapter 3, on the origin of the elements, is entirely concerned with nuclear reactions, and Chapters 4 and 5—covering the formation of the Earth and planets, and the distribution of elements on the Earth's surface, respectively—with the chemical behaviour of the elements. In the final chapter, on the distribution of isotopes, the two themes come together. This is a subtle, but interesting question which depends on both nuclear (especially radioactive) and chemical differences between the isotopes of an element.

Summary

The 90 chemical elements which occur naturally on Earth have a very wide range of abundance. Fe, O, Si, Mg, Ca, Al, S, and Ni each have an average abundance of more than 1 per cent, the remaining elements together making up only 1 per cent of the total mass. Several stable elements have abundances in the parts-per-billion range. The elements also show characteristic differences in distribution between different parts of the Earth: lithophiles concentrate in the oxide minerals of the mantle and crust, siderophiles in the metallic core; chalcophiles occur predominantly as sulphide minerals; atmophiles are present in volatile form in the atmosphere and oceans.

Analysis of meteorites, and spectroscopic observations on the Sun, give an estimate of the overall abundance of elements in the solar system. This is dominated by hydrogen and helium, the remaining elements making up only 1 per cent. From the study of distant stars and galaxies, it appears that composition of the solar system is fairly typical of the present universe. However, these observations also suggest that hydrogen and helium were the original constituents of the universe, and that the other elements have been produced since its beginning.

To understand the origin and abundance of the elements in the universe as a whole, we must study the nuclear processes by which elements can be transmuted into one another. On the other hand, the atypical composition of the Earth, and the distribution of elements within it, is a result of chemical fractionation.

Further reading

The development of chemistry, including the modern concept of an element and the periodic table, is described by Ihde (1964), and the story of the discovery of the elements by Weeks and Leicester (1968).

The book by Goldschmidt (1954) is a very influential work on geochemistry, by the originator of the important classification scheme described in the chapter. A good modern introduction to this subject is Mason and Moore (1982).

Tables of the abundances of the elements can be found from many sources. The data used in plotting the figures in this (and later) chapters are tabulated, with detailed references, in Appendix A at the end of the book. A good account of our present knowledge of the Earth's interior is Brown and Musset (1981), and very nicely illustrated discussions of many astronomical topics can be found in *The Cambridge Encyclopedia of Astronomy*. Hearnshaw (1986) gives an interesting historical account of the development of the spectroscopic methods which show the composition of the Sun and stars.

More references will be found after later chapters, in which the topics touched on here will be discussed in detail.

2
Nuclear and chemical properties

The previous chapter showed the importance of two very different properties of each element. The origin and overall abundance of elements in the universe are controlled by the stability and reactions of atomic nuclei. On the other hand, the way in which elements are distributed, between different parts of the solar system and within the Earth itself, is largely a chemical problem. In the present chapter we shall discuss briefly the basis of these properties. It is interesting to look at chemical and nuclear effects in a comparative way, and so to bring out the points of similarity and difference between them more clearly than in conventional treatments.

Mass, size, and energy

The nuclear model of the atom was proposed by Rutherford and his collaborators in 1911. The famous α particle scattering experiment of Geiger and Marsden showed that nearly all the mass of an atom is concentrated in a small region in the centre, the positively charged **nucleus** (see Fig. 2.1). Surrounding this is a cloud of much lighter negatively charged **electrons**, which orbit the nucleus rather like the planets around the Sun.

It is now known that nuclei are composed of positively charged **protons** and uncharged **neutrons**. These elementary particles, which have approximately the same mass, are collectively called **nucleons**. The nuclear charge, proportional to the number of protons, is balanced in a neutral atom by the same number of electrons; this is the **atomic number** (Z) of an element, which controls its chemical properties and its position in the periodic table (see Fig. 1.1 and Table 1.1). The mass of an atom, on the other hand, depends on the total number of nucleons. Each different nuclear species or **nuclide** can be specified by giving its atomic number Z and its **mass number** A, equal to:

$$A = Z + N \tag{2.1}$$

where N is the number of neutrons.

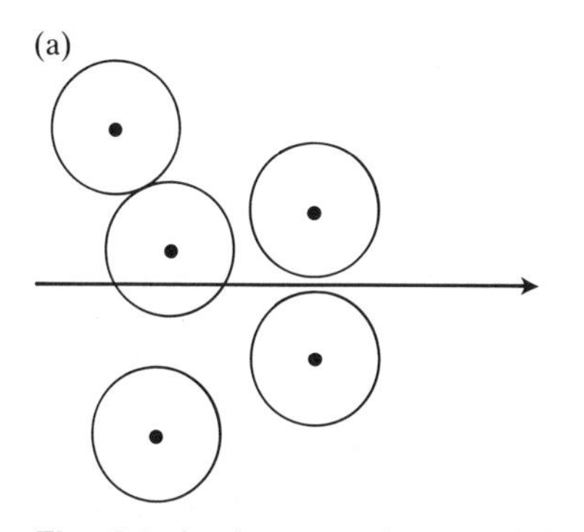

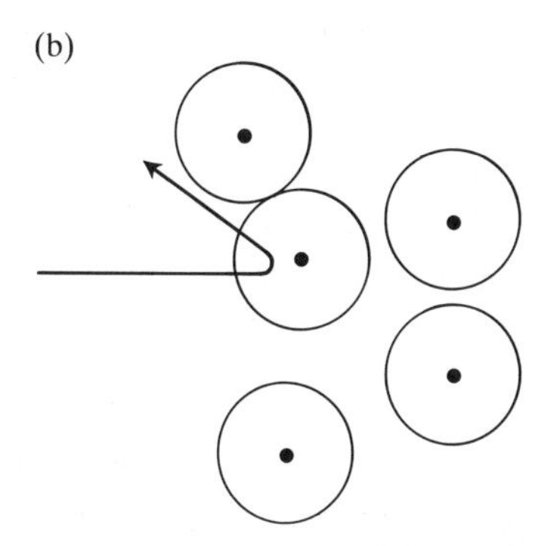

Fig. 2.1. In the experiment of Geiger and Marsden, a thin metal foil was bombarded with positively charged α particles. Most particles travelled through with very little deflection (a), but a few suffered very large deflections (b). This is only explicable if most of the mass of each atom is concentrated in a small positively charged nucleus at the centre.

Chemical measurements of the relative atomic mass of an element show only the average value for a very large number of atoms. With the invention of the **mass spectrometer** by Aston in 1921, it became possible to measure the masses of individual atoms, and it was then found that many elements consist of a mixture of atoms with different masses. These are known as **isotopes**, and have the same number of protons, but different numbers of neutrons, thus giving different mass numbers. For example magnesium, with $Z=12$, has three stable isotopes with $A=24$, 25, and 26, denoted ^{24}Mg, ^{25}Mg, and ^{26}Mg. In all naturally occurring magnesium they are present in very nearly the same proportions, 79.99, 10.00, and 11.01 per cent, respectively. The average mass number of natural Mg is therefore 24.320, which is quite close to its relative atomic mass (24.305). The discovery of isotopes partly solved the mystery of why the relative atomic masses of many elements are far from being exact multiples of a simple unit. Even so, small discrepancies remain, as the masses of atoms are not exactly proportional to their mass numbers. This is partly due to the presence of differing proportions of particles (protons, neutrons, and electrons) with different masses. As we shall see later, however, the precise masses of atoms also give important information about the stability of their nuclei.

A table of the natural isotopic compositions of the elements on the Earth is given in Appendix B at the end of this book. The chemical properties of different isotopes of an element are very similar, indeed indistinguishable for many purposes. However, the difference in mass does give small chemical differences, and sometimes much larger ones result from different nuclear properties. For some elements, natural samples from different sources do not have quite the same isotopic composition and this gives rise to small variations in relative atomic masses. The origin of these deviations and the interesting information that can be obtained from them are explained in Chapter 6.

For most elements the different isotopes are given the same chemical name and symbol, with the mass number shown as in the case of magnesium mentioned above. With hydrogen, however, its isotopes are referred to by different chemical names. By far the most abundant isotope is ^{1}H, with one proton and no neutrons in the nucleus. The heavier isotope with one neutron is called **deuterium** and written either ^{2}D or ^{2}H. There is also an unstable radioactive isotope, **tritium**, written ^{3}T or ^{3}H.

The overall size of an atom is determined by the radii of the electron orbits. Measurement of interatomic distances in crystals and molecules show that this size varies with the position of the element in the periodic table, but is generally of the order of 2×10^{-10} m (200 pm). By contrast the α particle scattering experiments in Rutherford's laboratory showed that the nucleus is much smaller, less than 10^{-14} m (=10 fm). More recent experiments, using the scattering of high energy electrons and protons, show that nuclei do have a measurable size. Their radii roughly follow the formula:

$$R = R_0 A^{1/3}, \tag{2.2}$$

where A is the mass number and R_0 a constant with a value of about 1.2 fm. Since volume is proportional to R^3, equation 2.2 expresses the fact that the **volume per nucleon** is approximately the same in all nuclei. It is tempting to think of a nucleus as a more-or-less close-packed collection of nucleons, with rather little free space between them. Atoms as a whole are quite different, as the electrons behave essentially as 'point charges' with no measurable size. Thus, the electrons in an atom are rather like molecules in a gas, where most of the volume is made up of free space. The nucleus, on the other hand, resembles more a drop of liquid, and although this picture must not be taken too literally, as it is now realized that nucleons have considerably more space to move around in than the idea suggests, it forms the basis for the simplest qualitative model of nuclear structure, described in the next section.

The very different sizes of atoms and nuclei lead immediately to an estimate of the different energy scales involved in chemical and nuclear reactions. According to the Coulomb law of electrostatics, the energy of interaction between two charges q_1 and q_2, separated by a distance r, is:

$$V = q_1 q_2/(4\pi\varepsilon_0 r). \tag{2.3}$$

With q_1 and q_2 equal to the electron or proton charge, and r equal to a typical atomic size 200 pm, we find an electrostatic energy of 1.2×10^{-18} J, or 7 eV. This is a typical value for the binding energy of an outer electron in an atom, and also gives an idea of the energies involved in making or breaking chemical bonds, a process which involves the redistribution of electrons between atoms. (In the **molar** units more familiar to chemists, the energy just quoted is about 700 kJ/mol.) For r equal to a nuclear size, on the other hand, we find an energy of about 1 MeV (or 100 GJ/mol). This is, indeed, the energy scale involved in nuclear reactions, where nucleons are redistributed between atoms, but apart from the different orders of magnitude there is another important contrast between the two estimates. Whereas the energy of **electrostatic attraction** between positive nuclei and negative electrons is the force responsible for chemical bonding, the only charged species in nuclei are the positive protons. The electrostatic energy is, therefore, **repulsive** in nature, and another, attractive force is required to balance it in order for nuclei to be stable. The nature of this is discussed later.

The very large energies involved in making or breaking up nuclei can be measured directly, but they are also apparent in a more subtle way, through the effect they have on atomic masses. We have mentioned already that the precise masses of atoms are not exact multiples of that of hydrogen. The agreed scale for the **atomic mass unit** (amu) is now based on the ^{12}C atom, which is defined to have a mass of exactly 12 amu. On this scale, the mass of a normal hydrogen atom (^{1}H) is 1.00727661 amu. The mass of a free neutron (1.00866520 amu) is also slightly greater than 1, and, at first sight, there is a puzzling discrepancy between the mass of a carbon atom and its constituent particles. If ^{12}C were made from six hydrogen atoms and six neutrons, there would be a mass equal to

$$(6\times1.00727661)+(6\times1.00866520)-12.00000000=0.095651 \text{ amu}$$

missing at the end. All atoms show this kind of discrepancy, having masses slightly less than that of their constituents.

This missing mass would be impossible to explain in classical physics, but it is consistent with a remarkable prediction made (before any experimental evidence was known) by Einstein's Special Theory of Relativity in 1905. Einstein's theory leads to the famous equation relating the mass of a body (m) to its energy (E), with the velocity of light (c):

$$E = mc^2. \tag{2.4}$$

As a result of this equation, Einstein was led to suppose that mass and energy are two aspects of the same 'mass-energy' quantity, and that changes in energy will give rise to corresponding changes in the measured mass. Because of the magnitude of c^2 (nearly 9×10^{16} J/kg), such mass differences are too small to measure for the energies involved in chemical or mechanical changes. With the much larger energies of nuclear processes however the mass changes are quite significant; indeed, the measurement of mass discrepancies such as that quoted for ^{12}C is often the best way of estimating an energy change. The fact that all nuclei have masses *less* than their constituent particles shows that the energy is also less. In fact, the mass loss gives a direct measurement of the energy released when a nucleus is formed from elementary particles; that is, its **binding energy**. Energies for nuclear processes are normally quoted in MeV, with the conversion factor:

$$c^2 = 931.5 \text{ MeV/amu}.$$

For example in ^{12}C, the total nuclear binding energy is

$$931.5 \times 0.0965 = 90.2 \text{ MeV}.$$

The precise measurement of atomic masses made possible by the mass spectrometer not only provided an important—and unexpected—verification of the Special Theory of Relativity. It also showed that nuclear reactions could give energies vastly greater than those obtainable by chemical means. This discovery has been quite influential, both in scientific and political terms, in the history of the twentieth century. As is well known, Einstein himself was initially involved in the proposal to build the first atomic weapons. In our present context, the realization that only nuclear processes could provide the energy source in stars helped to solve a major problem in astronomy and gave an important clue to the origin of the elements.

Nuclear forces: the 'liquid drop' model

We have just seen that an accurate measurement of the mass of an atom gives an estimate of the total binding energy of the protons and neutrons within the nucleus. It is often more instructive to divide this total energy by the mass number, to give the **binding energy per nucleon**. Figure 2.2 shows a plot of these values, for nuclides with masses extending up to 250. After a rapid (and irregular) rise at low A, the binding energy is roughly constant at around 7–8 MeV per nucleon. However, there is a slow decline after $A = 56$, and indeed ^{56}Fe is the most 'stable' of all nuclides, with the highest binding energy per nucleon. The form of Fig. 2.2 is very important in understanding the nuclear processes by which the elements are synthesized in stars. It can be seen that it is energetically favourable to form elements up to iron by the **fusion** of

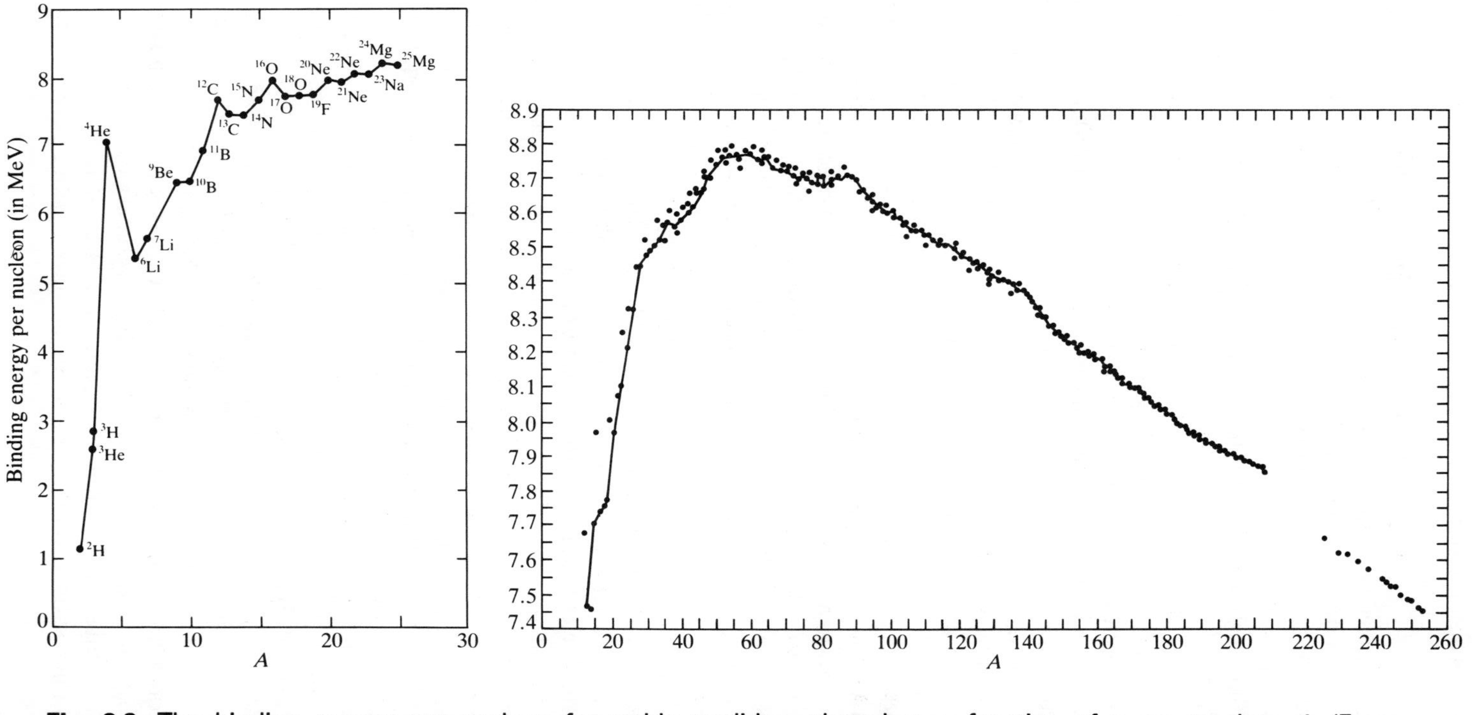

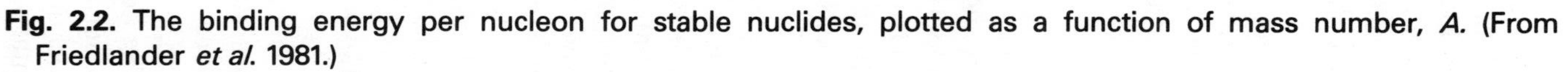

Fig. 2.2. The binding energy per nucleon for stable nuclides, plotted as a function of mass number, *A*. (From Friedlander *et al.* 1981.)

lighter nuclei. By far the most energy output comes from the first stage, the fusion of hydrogen atoms to form helium, and this is indeed the reaction which provides stars with an energy source for the major part of their lives. Elements beyond iron can be formed by fusion only **endothermically**, and it is, in fact, favourable to split heavier nuclei up into smaller fragments—the **fission** process used in nuclear weapons and power plants.

The stability of nuclei can be explained only if there is an attractive force operating between the nucleons, strong enough to overcome the repulsive electrostatic interactions between protons. This attractive nuclear force is known as the **strong interaction**, and it differs from the electrostatic force in two important ways. First, it operates equally between protons and neutrons. Secondly, it is a *short-range* force, its effect extending only over distances of around 2 fm or less. This limited range is in contrast to the 'inverse square law' electrostatic force, and explains why the strong interaction is important only at the nuclear size scale, and is unknown in classical 'macroscopic' physics. It is also crucial in understanding the overall shape of Fig. 2.2. If the attractive strong interaction were a long range foce, the binding energy per nucleon would continue to rise with increasing mass, and there would be no limit to the number of stable nuclides. As it is, the rough form of the binding-energy curve can be explained by the balance between the attractive short-range strong interaction, and the repulsive long-range electrostatic force. This is illustrated schematically in Fig. 2.3. The strong interaction operates only between nucleons that are near neighbours. Thus, for large nuclei, its contribution per nucleon is almost constant, since most nucleons are 'inside' the nucleus and have the maximum number of near neighbours (twelve, for an ideally 'close-packed' collection of nucleons of equal size). For light nuclei, the contribution is less, because most nucleons are on the 'surface' of the nucleus and have fewer neighbours. On the other hand, the repulsive electrostatic contribution rises more slowly with mass number, and continues to rise indefinitely, because the operation of this long-range force is not confined to near neighbours.

As shown in Fig. 2.3, the sum of the two contributions to the nuclear binding energy gives a result which qualitatively reproduces the experimental curve of Fig. 2.2. There is a rapid rise followed by a levelling off, and a slow fall where the long-range electrostatic repulsion starts to dominate. It is the dominance of the electrostatic force in heavy nuclei which, therefore, provides a limit to the number of stable elements.

The **liquid drop model** just described is useful as it provides a simple account of the main trends in nuclear binding energies. However, the model does have serious limitations, and cannot explain why the experimental values shown in Fig. 2.2 do not lie on a smooth curve. A more

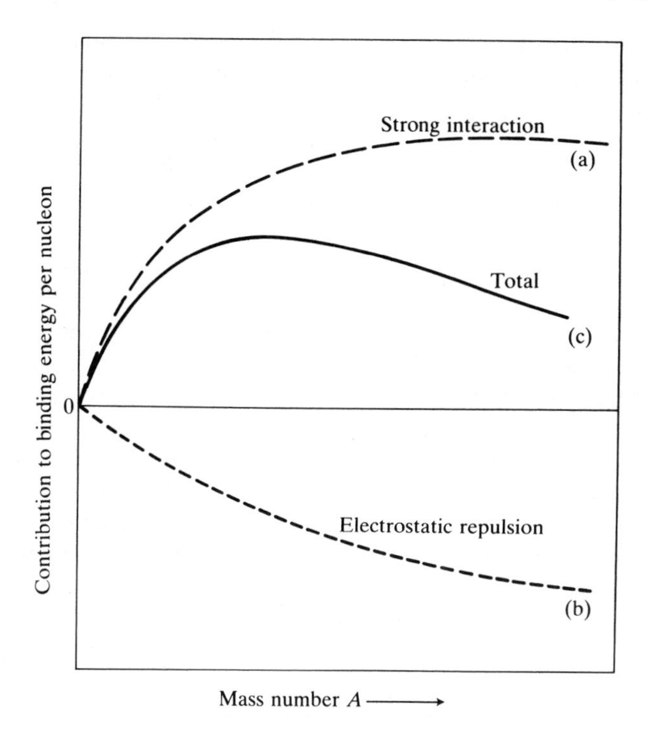

Fig. 2.3. Major contributions to the binding energy curve, Fig. 2.2. (a) Contribution of the attractive strong interaction. (b) Contribution of the electrostatic repulsion between protons. (c) Sum of (a) and (b).

satisfactory theory must take account of the way in which nucleons are able to move around in the nucleus. This theory depends on quantum mechanics and the resulting **shell model** is described briefly in the next section.

Before discussing the shell model, it is appropriate to consider another problem. The binding energy curve shows the stability of nuclei with respect to individual nucleons, but it also indicates that most nuclei are apparently *unstable* with respect to ^{56}Fe. What is it that inhibits the energetically favourable rearrangement of nucleons, by which lighter nuclei would be expected to fuse together, and heavier nuclei to split into smaller fragments? Both these processes do actually occur. As we have already said, the fusion of light nuclei is an essential step in the formation of heavier elements. However, fusion occurs naturally only at very high temperatures, where atoms have very high kinetic energies due to their thermal motion. This suggests that there is a large energy barrier which normally prevents the reaction. Many heavy nuclei also split up, especially by the radioactive process of α decay, in which a particle (a ^{4}He nucleus) is emitted. Such decays occur slowly, however, so that many unstable nuclei have life-times of billions of years or more. Once again, a large energy barrier seems to be involved. In fact, the barriers to both fusion and decay processes are similar, and are a direct consequence of the balance of the two kinds of force acting within the nucleus.

Figure 2.4 shows a schematic potential energy curve for two nuclei, as a function of the distance between them. At short distances the attractive strong interaction is dominant, leading to a state of lower energy than that of the separated constituents. However, beyond a distance given roughly by the sum of the nuclei sizes (a few fm), the strong interaction no longer operates, and the energy is then due to the long-range Coulomb repulsion, given by equation 2.3. The result is a large **Coulomb barrier**, representing the electrostatic repulsion that must be overcome before the nuclei get close enough for the strong interaction to operate. Using equation 2.2 for the nuclear size gives an estimate for the barrier height:

$$B = 1.2\, Z_1 Z_2 / (A_1^{1/3} + A_2^{1/3}) \text{ MeV}. \qquad (2.5)$$

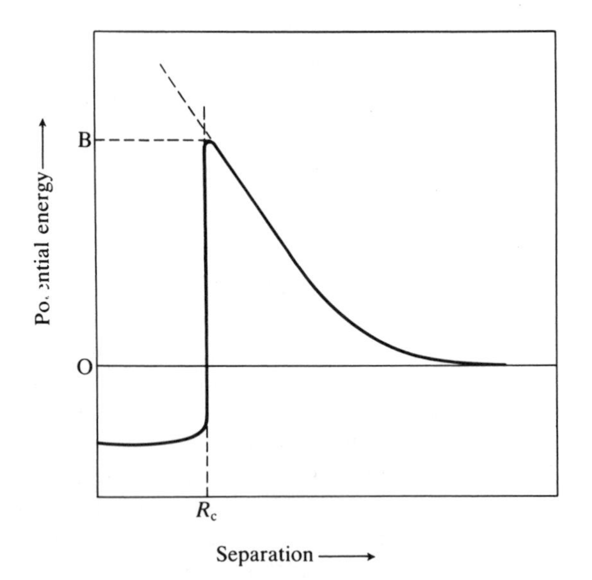

Fig. 2.4 The Coulomb barrier for nuclear reactions and decay. The potential energy of two nuclei is plotted as a function of their separation. At distances larger than R_c only the Coulomb repulsion is effective; the strong interaction comes into play at shorter ranges.

This shows very clearly why high energies are required for fusion reactions, as it predicts that light nuclei would require kinetic energies of the order of 1 MeV or more to get close enough for the reaction to occur.

The Coulomb barrier argument applies equally to the radioactive decay of heavy nuclei. For example, an α particle emerging from ^{210}Bi will experience a repulsive Coulomb potential of about 28 MeV at the point where it passes out of range of the strong interaction. Once again, we can understand the difficulty of this process, but these estimates of the barrier height pose a problem of a different kind. The actual energy of α particles emitted in the decay of ^{210}Bi (determined by the difference of binding energy of the initial nucleus and the two products) is about 6 MeV. An α particle with this energy should not be able to cross the barrier at all. A similar difficulty arises in fusion reactions, which are found to take place between nuclei with insufficient energy to pass over the predicted Coulomb barriers.

The problem of how an α particle can pass the Coulomb barrier was solved independently by Gamow and by Gurney and Teller in 1928, and their explanation formed the first application of an important prediction of quantum mechanics.

According to classical physics, a particle cannot cross a region where the potential energy (V) exceeds its total energy (E). However, solutions of Schrödinger's equation show that light particles do have a small probability of passing through a 'classically forbidden' region. This process is known as **tunnelling**, and its probability depends on the width of the barrier and the 'energy deficit' ($V-E$). Calculations give the following approximate formula for the probability of tunnelling in nuclear reactions or radioactive decay:

$$P = \exp[-Z_1 Z_2 e^2/(\hbar \varepsilon_0 v)], \tag{2.6}$$

where v is the relative velocity of the separated nuclei, related to the square root of the incident energy (in a nuclear reaction), or of the energy release (in decay processes). It is typical of the quantum theory that this formula gives only a probability. Radioactive decay is a fundamentally *statistical* process, occuring at a moment which is totally unpredictable for any given nucleus. All that can be estimated is the average rate of decay, for example, the number of nuclear disintegrations per unit time from a large sample. Decay rates are usually expressed as a **half-life**, meaning the time during which half the nuclei in a given sample are likely to decay. Such half-lives are inversely proportional to the decay probability given in equation 2.6. Barrier tunnelling rates for radioactive decay are essential independent of variables such as temperature, pressure, and the chemical state of a given atom. Half-lives are, therefore, unaffected by these parameters.

Equation 2.6 predicts a very low probability of tunnelling for small values of the velocity v. Thus, the collision energy in a nuclear reaction must be quite high for it to occur at a significant rate. In α decay, low decay energies also lead to low tunnelling probabilities and, hence, to long half-lives. A detailed consideration of binding energies suggests that all nuclei with $A > 150$ ought to be unstable to α decay, which implies that all elements beyond the lanthanide neodymium ($Z = 60$) should be radioactive. In fact many elements with $Z > 60$ are known to be α emitters, but with extremely long half-lives, often more than 10^{15} years. This is because the energy of the emitted α particle is so low that barrier tunnelling has a very small probability. It is only for elements beyond bismuth that the energy release becomes great enough for decay to limit the stability of the elements significantly. This process continues for the heavier elements as the binding energy per nucleon falls increasingly steeply and the α decay energies increase. For $Z = 92$, the longest lived isotope ^{238}U has an α decay half-life of 4.51×10^9 years; at $Z = 96$ (^{247}Cm) this has fallen to 4×10^7 years, and by $Z = 100$ (^{252}Fm) to 21 hours.

Nuclei can also decay by another route, that of **spontaneous fission** into two roughly equal fragments. Such a process is, in fact, energetically more favourable than α decay, but it normally occurs rather rarely, at least up to the very heavy nuclide ^{248}Cm. This can be explained by using equation 2.6. The term Z_1Z_2 (representing the barrier height) is larger for fission than for α decay; also for a given energy release the velocity v will be lower for heavier fragments. Thus, the probability of barrier tunnelling is much lower for fission. For very heavy elements this formula breaks down, however, as their nuclei are actually far from spherical. Beyond curium, spontaneous fission is commonly observed, and it is this process, together with α decay, that limits the life-times of very heavy elements.

The shell theory of atoms and nuclei

Although many essential features of nuclear stability can be accounted for by the 'classical' liquid drop model described in the previous section, some important anomalies remain. We have commented on the fact that nuclear binding energies do not fall on the smooth curve predicted by the model. These deviations can be seen more clearly by plotting, not the average binding energy as in Fig. 2.2, but the difference in binding energy between successive nuclides. The **separation energies** of protons or neutrons are the energies required to remove one nucleon from a nucleus, and are analogous to the ionization energies of electrons in atoms discussed below. Figure 2.5 gives examples of these, showing in (a) the neutron separation energies for isotopes of Br ($Z = 35$) and Kr ($Z = 36$), and in (b) the proton separation energies for a series of nuclides with a constant neutron number $N = 50$. Both plots show the same general features.

1. There is a regular alternation, separation energies being larger for nuclides with an **even** number of the corresponding nucleon than those with an **odd** number.
2. There is a discontinuity in the average trend of separation energies, occurring after the nuclide with $N = 50$ in Fig. 2.5(a), and after that with $Z = 50$ in Fig. 2.5(b).

The even–odd alternation of energies is shown throughout the series of known nuclides, and as discussed in Chapter 1 (see Fig. 1.7 on p. 17) is reflected in the pattern of elemental abundances found in the universe. Nuclei with even numbers of protons and/or neutrons are apparently more stable than those with a corresponding odd number. The drop in stability after 50 protons or neutrons is also found with some other values of Z and N. These are 2, 8, 20, 28, 50, 82, and 126, and are known as the **magic numbers**. Nuclei with a magic number of either protons or neutrons appear to be more stable than others. This pattern is also reflected to some extent in the abundances of the elements: for example, Fig. 1.7 shows a broad peak in abundance between tin and barium, associated with the magic numbers $Z = 50$ (isotopes of tin) and $N = 82$ (especially ^{138}Ba).

The nucleon separation energies have their analogue in the **ionization energies** of electrons from atoms. Since the chemical properties of an element depend to a large extent on the behaviour of the least-tightly-bound electrons in the atom, such ionization energies play an important part in rationalizing chemical trends. Figure 2.6 shows a plot of first ionization energies against the positions of elements in the periodic table. This emphasizes the strongly periodic nature of electronic energy trends. There is a general rise with increasing group number, followed by a marked drop between groups 18 (the noble gases) and 1 (the alkali metals). The drop is similar to the magic number effect in nuclei, although it is more dramatic in the atomic case.

The patterns illustrated in Figs 2.5 and 2.6 cannot be explained by classical models like the liquid drop, and are manifestations of the quantum mechanical **shell structure** of atoms and nuclei. The shell theory was first developed to describe the motion of electrons in atoms. Although Rutherford's 'planetary system' picture of the atom is very appealing, it suffers from a major problem. In classical physics, the electrons orbiting the nucleus should lose energy in the form of radiation

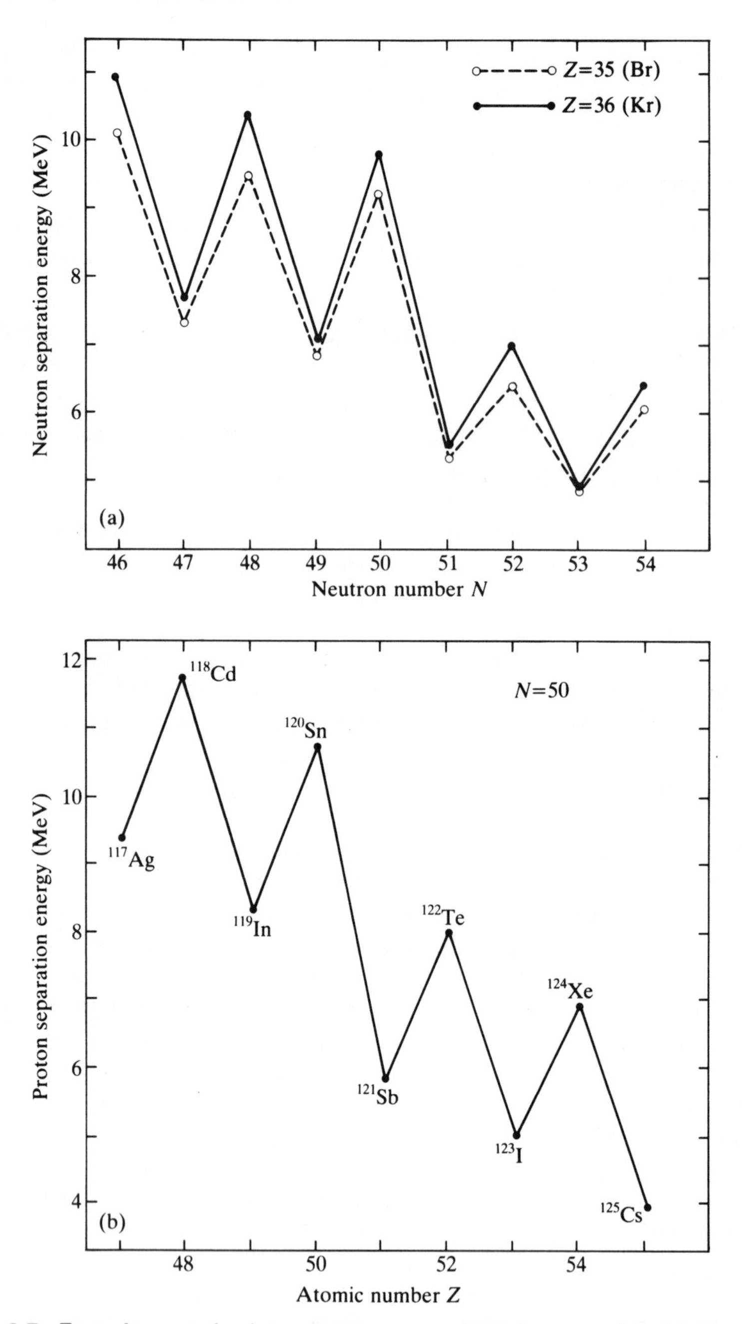

Fig. 2.5. Energies required to separate a nucleon from nuclei. (a) Neutron separation energies for isotopes of Br ($Z=35$) and Kr ($Z=36$) plotted against neutron number N. (b) Proton separation energies for nuclides with $N=50$, plotted against proton number Z.

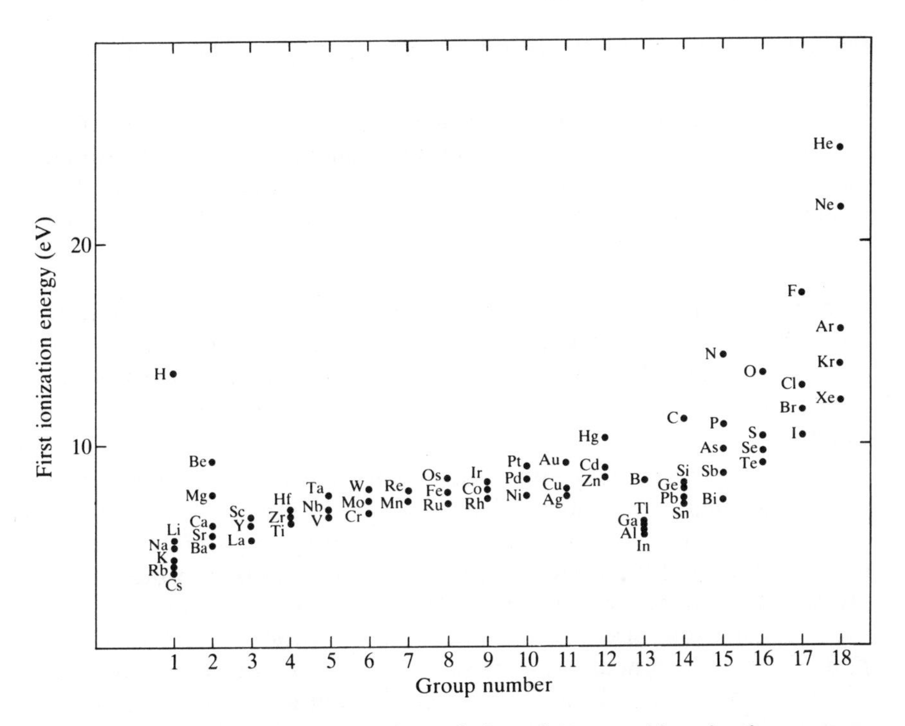

Fig. 2.6. First ionization energies of the elements (that is the energy required to remove one electron from a neutral atom) plotted against their position (group number) in the periodic table.

and hence spiral rapidly into the nucleus. In 1913 Niehls Bohr suggested that some of the recently developed ideas of the quantum theory could be used to overcome this problem. He proposed that only certain orbits with well-defined energies are allowed, so that electrons cannot lose energy in a continuous manner as predicted by classical physics. A more complete explanation of these allowed orbits was provided by Schrödinger's wave equation in 1926. Indeed, the explanation of atomic structure and the periodic table was the first major triumph of the 'New Quantum Theory'. The same ideas were applied to nuclear structure in the years around 1950.

Schrödinger's theory is highly mathematical, but the essential results are as follows.

1. The motion of microscopic particles such as electrons is not described by well-defined trajectories as in classical physics, but by functions known as **orbitals** which determine the probability of finding a particle at a particular place. Only certain energies give acceptable

orbitals as solutions to Schrödinger's equation. Thus, energy is **quantized**.

2. The allowed energies of the orbitals depend on the potential field in which particles are moving. However, in a spherical system each orbital may be characterized by the amount of rotation or **angular momentum** of the particle. Orbitals are denoted (for historical reasons) by the letters

$$s, p, d, f, g, h, i, \ldots$$

associated with 0, 1, 2, 3, ... units of angular momentum, respectively. Furthermore, each level with l units of angular momentum has $(2l+1)$ separate orbitals, all with the same energy. Thus, there is one s orbital with a given energy, 3 p orbitals with the same energy, 5 d orbitals, and so on.

3. Electrons and many other elementary particles also possess an intrinsic angular momentum, called **spin**. (This is rather like the rotation of a planet about its own axis, although the analogy must not be pressed too far.) The spin of an electron or a nucleon can point in one of two directions. The **Pauli exclusion principle** states that only two electrons, with opposite spin directions, can occupy each orbital. It follows that s, p, d, ... levels can hold a total of 2, 6, 10, ... electrons, respectively. In principle, the same rules apply to nucleons, but as we shall see below, their use is a little more complicated.

Figure 2.7 shows schematically the sequence of orbital levels predicted for electrons in atoms, where the potential field comes from the Coulomb attraction to the nucleus (equation 2.3), partly offset by repulsion from other electrons. Orbitals are denoted $1s$, $2s$, $2p$, ... etc., giving first their **principal quantum number** and then the letter showing their angular momentum. The sequence of orbital energies in Fig. 2.7 shows clearly how the periodic table of elements (Fig. 1.1) is associated with the shell structure of the atoms. The noble gas atoms of group 18 have completely filled orbital groups or **shells**, with an appreciable energy gap to the next available orbital. This higher orbital is then occupied by one electron in group 1 elements. Elements in the same group have the same number of electrons in their outermost shell, which explains their chemical similarities. We can see how more elements are accommodated in the later periods, by the filling of the d shells (corresponding to the 10 transition elements) from the fourth period onwards, and of the f shells (giving the 14 lanthanide and actinide elements) starting in the sixth period.

The alkali metals of group 1 have low ionization energies because of their electron in a high energy (and, therefore, relatively less stable) orbital. Along each period the increasing nuclear charge attracts the

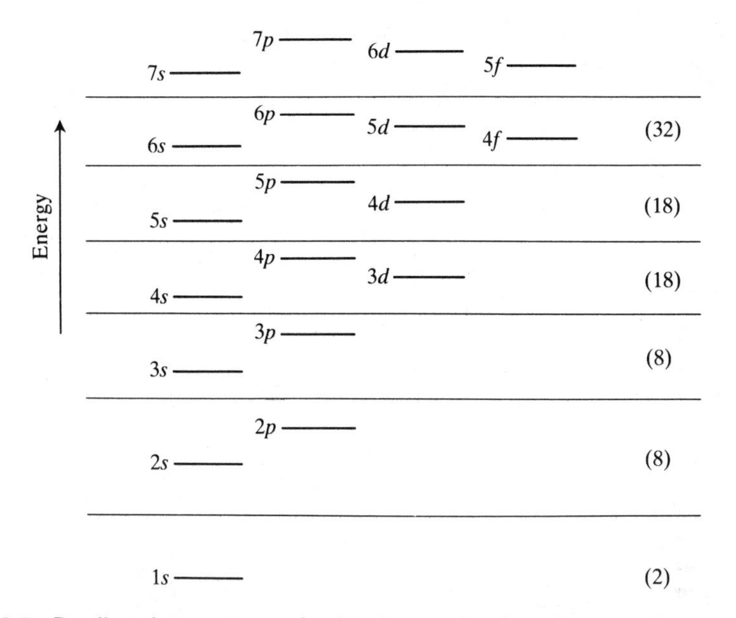

Fig. 2.7. Predicted sequence of orbital energies for electrons in atoms. *s* levels can hold two electrons each; *p*, *d*, and *f* levels, respectively, 6, 10, and 14. The diagram indicates the number of electrons which can be accommodated in each successive row in the periodic table, ending with the filled-shell noble gas elements in group 18.

electrons in a given shell more strongly. The ionization energy of successive elements therefore tends to rise, although breaks in this general trend are associated with the filling of 'sub-shells', the individual orbital levels making up the complete shell.

Whereas electrons in an atom are attracted to the central nucleus, the particles within the nucleus itself are attracted to each other and not to anything else at the centre. Nevertheless, each nucleon does move in a potential well, determined by the strong interaction to the other nucleons (partly offset in the case of protons by their repulsive Coulomb interactions). As with the 'Coulomb well' for electrons, solution of Schrödinger's equation gives a succession of allowed orbital energies for nucleons, illustrated in Fig. 2.8. There are a number of important differences from the orbital structure of atoms. In the first place, the shape of the well leads to a different sequence of orbital levels. By convention (and this has no fundamental significance) they are also labelled differently, the lowest *p* and *d* orbitals in nuclei, for example, being called 1*p* and 1*d*, not 2*p* and 3*d* as in atoms. The second important difference is the influence of **spin-orbit coupling** in nuclei.

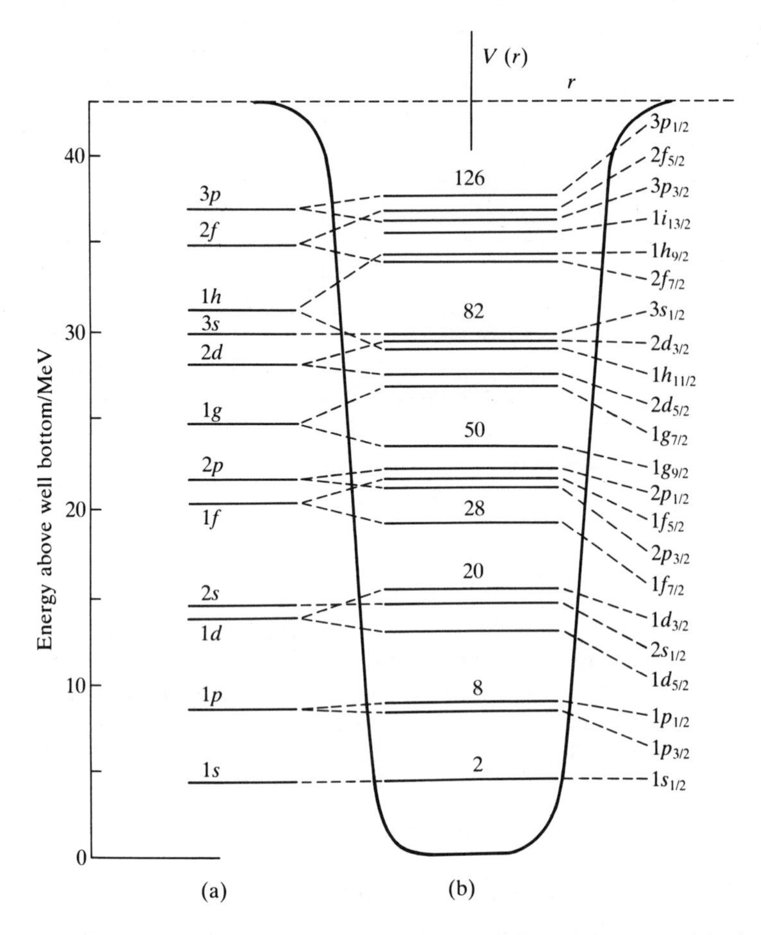

Fig. 2.8. The approximate sequence of predicted energy levels for nucleons within a potential well. (a) Without spin-orbit coupling. (b) With spin-orbit coupling included. The magic number configurations are indicated. (From Burcham 1979.)

This phenomenon does occur in atoms, where it arises from the interaction of the spin of the electron with the magnetic field produced by its orbital motion. However, it is a small effect for electrons, and does not perturb their energies sufficiently to upset the shell structure of the periodic table. Spin-orbit interaction is much stronger in nuclei and occurs because the strong interaction between nucleons depends on their relative spin direction. If it is ignored, the correct magic numbers cannot be obtained. As shown in Fig. 1.8, the consequence of spin-orbit coupling is to split all orbitals with $l > 0$ (that is, all except s orbitals) into the two levels, with total angular momentum (j) given by the values:

$$j = l + 1/2 \quad \text{and} \quad l - 1/2.$$

The higher j value gives the lower energy level. (This is also different from the situation with electrons.)

Like electrons, nucleons obey the exclusion principle, but the rules for applying this must be modified to take account of spin-orbit coupling. The result is that a level with total orbital angular momemtum j can hold $2j + 1$ nucleons of the same kind. With this principle, a suitable sequence of energy levels can explain the existence of the magic numbers. As shown in Fig. 2.8, these occur whenever there is a filled shell of nucleons, analgous to the electron configurations found for the noble gases.

The strong interaction provides a potential well, with a corresponding set of orbital levels, for both protons and neutrons. The depths of these wells will be different, because protons also experience the repulsive electrostatic forces. However, the overall result is that a magic number of either type of nucleon forms a closed shell, like the inert gas configuration of an atom. Further nucleons must occupy orbitals of higher energy and so are easier to remove. It can be seen also from Fig. 2.8 that smaller effects might be expected with some other numbers of nucleons, corresponding to the filling of sub-shells.

The regular alternation of nucleon separation energies (Fig. 2.5) has no simple parallel in the atomic ionization energy plot (Fig. 2.6). The reason for this alternation is that two protons or neutrons in an unfilled shell can form a stable **paired** state, with their spins pointing in opposite directions. It is easier to remove an odd **unpaired** proton or neutron than one from a paired configuration. The effect is analogous to the pairing of electron spins in chemical bonds, and to a chemist it may not seem surprising that paired spin states of nucleons are favourable. After all, the usual situation in molecules is to have pairs of electrons with opposite spin. It does *not* happen with electrons in isolated atoms, however, and so ionization potentials do not show the same even–odd alternation effect displayed by nuclei. Although paired electrons occur in closed-shell atoms (where spin pairing is required by the exclusion principle), in other cases the lowest energy state is formed by arranging the electrons with their spin directions **parallel**. The influence of spin is, in fact, a rather subtle one and arises because the relative spin directions of two identical particles affect their spatial distribution. Particles with paired spins move closer together than simple classical arguments would predict, and when the spins are parallel they tend to keep apart. In atoms, it is energetically more favourable for electrons to keep away from each other, as this reduces their electrostatic repulsion. Thus, parallel spins are preferred. In chemical bonds, on the other hand, it is generally better for two electrons to come closer together, because by doing this they can both occupy the region between two atoms where their potential energy is favourable. The situation in nuclei is different from that in atoms, since the attractive strong interaction between a pair of nucleons dominates over the repulsive electrostatic force. It is, therefore, favourable for nucleons to pair with opposed spins, so that they come closer together. The pairing of spins in nuclei resembles that of electrons in molecules, although the reasons are quite different in the two cases.

The three stages in our discussion—the classical liquid-drop model, the shell model, and the pairing effect—together provide a reasonably satisfactory explanation of the main features of nuclear structure, and in later sections we shall see some of the influences of the magic number and pairing effects. Before doing this, however, the next section looks at some aspects of the chemical behaviour of the elements and how this is influenced by the electronic shell structure of the atoms.

Chemical trends in the periodic table

Chemical bonding is associated with a redistribution of electrons between atoms. Only electrons in the outermost orbitals are affected, those in inner filled shells being too tightly bound to take part. Therefore, we can understand why elements in the same group in the periodic table, with the same number of outer electrons, have similar chemical behaviour.

In **covalent** bonds, electrons are **shared** between neighbouring atoms. This is generally associated with a pairing of electron spins, so that the electrons concerned can come together in the region where they experience the favourable electrostatic potential of two nuclei simultaneously. The simplest covalent bond is that in H_2, which is probably the commonest molecule in the universe. On the Earth's surface, examples of covalent bonding are found in small molecules such as O_2, N_2, CO_2, and H_2O, which make up the major part of the atmosphere and oceans. Carbon also forces many series of compounds containing covalent C–C bonds: the other elements present may be hydrogen, oxygen, and nitrogen, and these compounds are important not only as hydrocarbon fuels, but also as the molecules of life. (We shall see in Chapter 4 that many carbon-containing molecules are also present in space.)

Covalent sharing of electrons is to some extent a feature of all chemical bonds, but truly equal sharing is only present in bonds between atoms of the same element. In other cases, the electron distribution has some asymmetry, being polarized towards one atom rather than the other. The trend in atomic ionization energies (Fig. 2.6) shows that the binding energy of the outermost electrons generally increases with the group number across the periodic table. The same trend is shown in chemical bonding. Elements on the left-hand side have relatively weakly bound electrons, which can be more-or-less completely lost, to form positive **ions** such as Na^+ and Mg^{2+}. These elements are known as **electropositive**, and their most stable compounds are formed with strongly **electronegative** elements on the right-hand side of the table. Elements in the later groups (apart from the noble gases of group 18, which have closed shells and display very little tendency to chemical bond formation under normal conditions) have the ability to gain extra electrons and so form negative ions, such as O^{2-} and F^-. Thus, we have the most familiar compounds of the lithosphere, the oxides of electropositive elements. The stability of these compounds is largely due to the electrostatic energy of attraction between positive and negative ions, which more than compensates for the energy input required to redistribute the electrons. They are normally solids, because the greatest electrostatic stability is achieved in crystal structures where positive ions are surrounded by several negative ones, and vice versa.

The **ionic model** of bonding just described has many limitations. The elements of the later transition and the post-transition groups are ones which, as we have seen earlier (see Fig. 1.5 on p. 13) are not found on the Earth so frequently as oxides. These elements have higher ionization energies than those of the early groups, and are not so strongly electropositive in character. Many of them occur in combination with sulphur, which is not so markedly electronegative as oxygen. Almost certainly, the ionic picture of bonding is less appropriate here, and most sulphide minerals should be thought of as having a large component of covalent bonding, accompanying a lesser degree of electron transfer. Even with oxides, the ionic model should not be taken too literally. For example, the bonding between silicon and oxygen has a large degree of covalency. Nevertheless, it is common in geochemistry to use the ionic picture to interpret the formulae and structures of a wide variety of oxide minerals, including silicates. We can assign the **formal** charge Si^{4+} to silicon, in combination with O^{2-}, as a way of rationalizing formulae in terms of an overall charge balance. Thus, we have SiO_2 (**quartz**), $MgSiO_3$ (**olivine**, containing Mg^{2+}), $KAlSi_3O_8$ (**potassium feldspar**, with K^+ and Al^{3+}), the sum of the ionic charges in each compound being zero. Formal ionic charges are also known as **oxidation states**, which is, in many ways, a preferable term, as it does not imply a 'real' ionic charge.

Some of the features of silicate minerals will be described in more detail in Chapter 5. We shall see there that, as well as the oxidation states which control the **stoichiometry** or relative numbers of different atoms present in a compound, the *sizes* of the different ions also play an important part in influencing the distributions of elements in a mineral. In the first place relative sizes largely determine the **co-ordination number** of atoms (that is, the number of atoms of another type which surround them in the crystal) and, hence, control the type of crystal structure formed by a given combination of elements. Secondly, ions of similar size can easily replace each other within the same mineral structure. For example, olivine generally contains a proportion of iron, by the partial replacement of Mg^{2+} by Fe^{2+}. Like ionic charges, **ionic radii** are slightly nebulous quantities. They are derived by splitting the experimental distances between atoms in crystals, in a somewhat arbitrary manner, into the sum of radii of the two ions. As well as the problem of this arbitrary division, there is also the difficulty that the distances between atoms of two given elements is found to vary somewhat, especially if the co-ordination number changes. In spite of these difficulties, ionic radii are useful quantities in geochemistry, especially for the lithophilic elements which form oxide minerals. Figure 2.9 shows a selection, giving elements with their most common formal charges or oxidation states, and their radii.

At a simple level, the ionic radii in Fig. 2.9 can be used to interpret some major trends in mineral structures. Ions of the smallest radii (50 pm or less) are most commonly found in structures where they have four oxide neighbours; larger ions (70–80 pm) tend to have octahedral co-ordination, with six neighbours; and positive ions of 100 pm or larger generally have higher co-ordination numbers, eight or more.

We can see from Fig. 2.9 that many oxidation states are related in an obvious way to the position of the elements in the periodic table. For the early groups, the commonest oxidation state is the same as the group number, with all the electrons in the outermost shell involved in bonding. This is no longer true for elements in later groups, as it is not energetically favourable to involve all electrons. According to the ionic model, the stable oxidation states are those which give the most favourable balance between the energies of ionization, to form a positive ion, and the electrostatic stabilization obtained in the crystal lattice. Sometimes, this results in quite an even balance between two or more different values: for example, iron occurs as both Fe^{2+} and Fe^{3+}, the higher oxidation state being predominant in contact with free oxygen at the Earth's surface, whereas the 2^+ state is common in the deeper rocks of the lower crust and the mantle. The sizes of ions also show some obvious regularities. Generally speaking, ionic radius increases down a

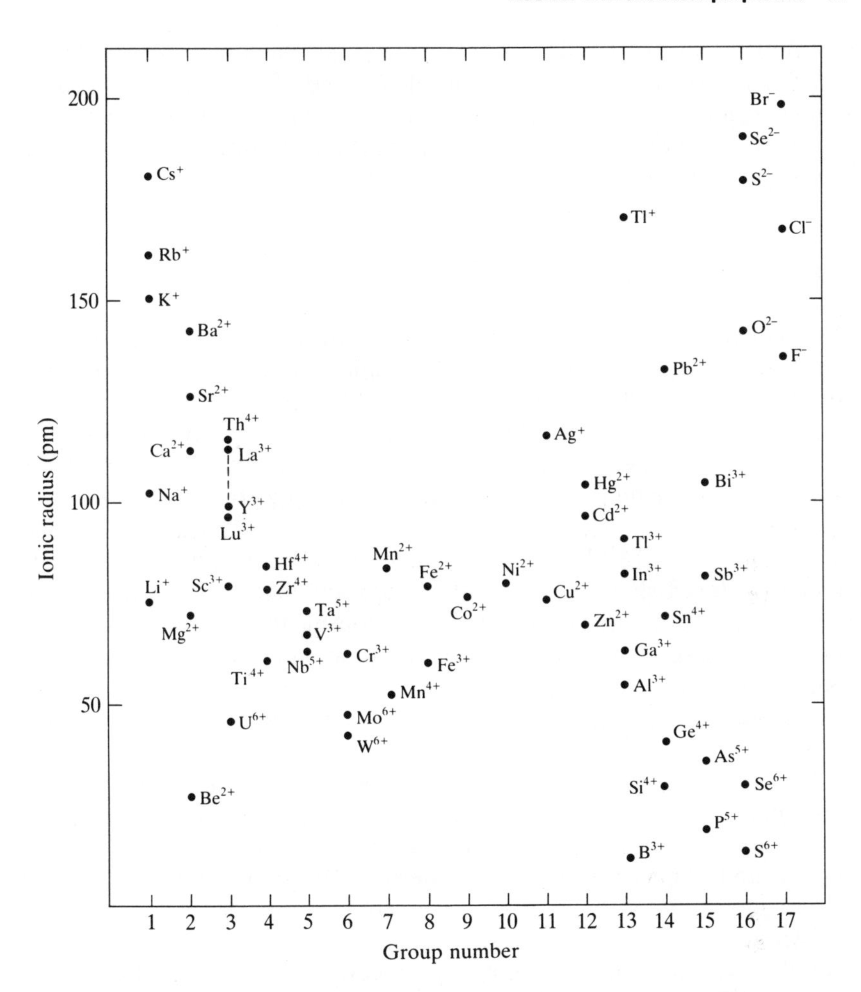

Fig. 2.9. Formal charges (oxidation states) and ionic radii of the elements, especially the lithophiles common in the Earth's crust. The values are most appropriate to oxides, with elements having their most common co-ordination number. (Data from Shannon and Prewitt 1969, 1970.)

group, where atoms become larger because of the filled inner electron shells. Size decreases with increasing charge (for example, along the series K^+, Ca^{2+}, Sc^{3+}, Ti^{4+}). These ions have the same number of electrons and the decrease of size is a result of the increasing nuclear charge which attracts the electrons more strongly. We shall see that a consequence of these trends is that the greatest geochemical similarities are not always shown by elements in the same group. For example, Na^+

can replace the similar sized Ca^{2+} in minerals much more easily than it can the larger K^+. In this case the difference of ionic charges can be compensated by another replacement (for example, of Si^{4+} by Al^{3+}). The way in which these effects influence the distribution of elements in the Earth's crust will be discussed in Chapter 5.

β decay and the 'line of stability'

We shall return now to another aspect of nuclear stability, and look at some of the effects that determine the stable isotopes of a given element. Stable nuclides of low mass contain roughly equal numbers of protons and neutrons. Typical examples are ^4He (where $Z=N=2$) and ^{16}O ($Z=N=8$). As the mass increases, however, it is found that the stable isotopes of an element tend to have an increasingly greater proportion of neutrons. For example, one of the heaviest stable nuclides is ^{208}Pb, with $Z=82$ and $N=126$. (Like the cases quoted above, this is an example of a **double-closed-shell** nucleus, with a magic number of both protons and neutrons.) The trend is illustrated by plotting the naturally occurring nuclei on a diagram that shows their proton and neutron number. Known as a **Segré chart**, this is shown in Fig. 2.10. Beyond $Z=20$ (^{40}Ca) the stable nuclei follow a path which diverges progressively from the line $Z=N$.

The general features of the 'line of stability' on the Z–N diagram can be explained using the ideas discussed previously. If the electrostatic force could be neglected, then all stable nuclei would be predicted to have Z and N nearly equal. This is because the strong interaction alone would give potential wells of the same depth for the two types of particle. If there were more neutrons than protons, for example, then the neutrons would have to occupy orbitals of higher energy in the shell-model diagram. This would be unfavourable, as the nucleus could find a state of lower energy by converting some of the excess neutrons into protons. For light nuclei, the electrostatic interaction is not very important compared with the strong interaction, and they do, indeed, have Z and N nearly equal. As our discussion of binding energies showed, however, the long-range electrostatic force becomes more significant in heavier nuclei. The repulsion between protons raises their potential energy and greater stability can now be achieved by changing some of them into neutrons. As the mass number increases, so the electrostatic interaction becomes more important, and it is favourable to convert a correspondingly greater proportion of protons into neutrons. Thus, the line of stability in Fig. 2.10, like the trends in binding energy discussed earlier, can be

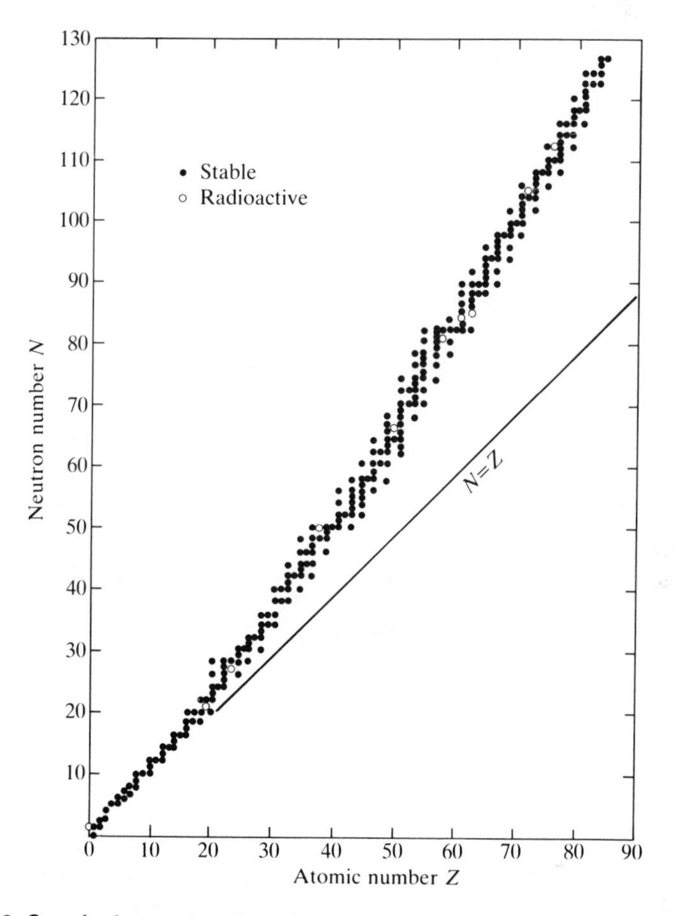

Fig. 2.10 Segré chart, showing the neutron number (N) and charge (Z) of nuclides stable to β decay. (From Burcham 1979.)

qualitatively explained as consequence of the balance of the two main forces—attractive and repulsive—that operate in the nucleus.

Nuclides lying off the line of stability are subject to β decay processes, by which neutrons are changed into protons or vice versa. Since this process does not change the mass number A, it is instructive to consider a series of nuclides with the same value of A. Two such series, one with odd A and the other even, are shown in Fig. 2.11. In each case, the total energy of the nuclides is plotted against the charge Z. A nuclide higher up on the plot can decay to one lower in energy. There are, in fact, three possible types of β decay.

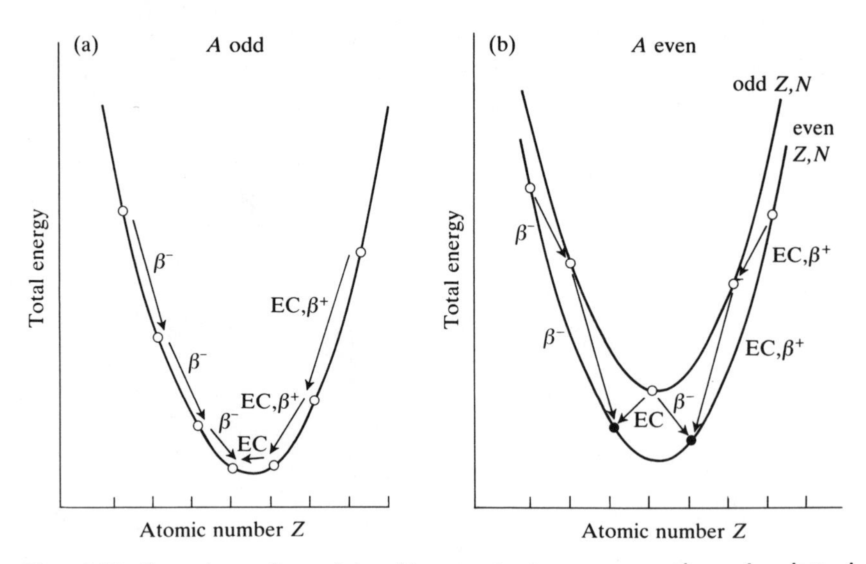

Fig. 2.11. Energies of nuclei with constant mass number A, plotted against the charge Z. (a) Odd A. (b) Typical case for even A, showing the alternation of energies. Unstable nuclei are marked (○) and the decay processes indicated: stable nuclei are marked (●).

1. Emission of a normal (negatively charged) electron changes a neutron into a proton and, hence, a nucleus with charge Z into one with $Z+1$. This is $\boldsymbol{\beta^-}$ **decay**.
2. A proton may be changed into a neutron, so that the nuclear charge goes from Z to $Z-1$, by the process of **electron capture** (EC), usually from the innermost ($1s$) orbital of the starting atom.
3. The same result as in case (2) may be achieved by emitting a **positron**—that is, the positively charged **antiparticle** of an electron. This is known as $\boldsymbol{\beta^+}$ **decay**.

Since the mass—and hence, by equation 2.4, the energy—of a neutron is slightly greater than that of a hydrogen atom, free neutrons are themselves unstable, and undergo β^- decay with a half-life of about 11 minutes.

Looking at Fig. 2.11, it can be seen that for an odd mass number all the energies lie on a single, smooth, roughly parabolic curve. In this case, we expect the end result of β decay and EC processes to give a single stable nuclide, at the lowest point of the curve, for each mass number. On the other hand, the even mass isobars alternate between two curves. This is because such nuclides have either an even number of both protons and neutrons or an odd number of both. The pairing between nucleons discussed above makes the even–even nuclei more stable, and they lie on a curve of lower energy than the odd–odd nuclei. In the most common situation, shown in Fig. 2.11 (b), β decay leads down to two nuclides, each with even Z and N. In principle, it is possible for the higher energy of these to decay to the other by a **double β decay**, where Z changes by two units simultaneously. Although this process has recently been observed, it is extremely slow and can normally be ignored. Thus, the commonest situation with even mass number is that there are two stable nuclides with the same A, each with even Z and N. It is only in a few cases with light nuclei (for example ^{14}N) that stable nuclides have odd Z and odd N. This is because for light nuclei the energies increase more steeply with changing Z, so that the lowest energy nuclide can lie on the *upper* of the two curves in Fig. 2.11(b). Thus, for example, ^{14}N is more stable than either ^{14}B or ^{14}C. For heavy nuclei, the energy curves are not so steep, and it is sometimes possible to find three or more stable even–even nuclei with a given mass number.

The discussion just given goes some way to explaining the pattern of stable isotopes, and the extremely wide diversity found between different elements. For example, elements with odd atomic number Z have only one or two stable isotopes, usually of odd mass number, where N is even. Elements with even Z may have several stable isotopes, those with even mass (and hence N) being more common. However, many of the details are not easy to 'explain' in any simple manner, and must reflect irregularities in the general trend of binding energies caused by shell filling and other effects. Elements where Z is a magic number can have an unusually large number of stable isotopes. Similarly, there may be a large number of stable nuclides with the same neutron number N, when this is a magic number.

Our discussion shows that at least one nuclide of each mass number should be stable to β decay.* This provides no guarantee, however, that every element should have a stable isotope. In fact, two elements in the middle of periodic table, technetium with $Z=43$, and promethium with $Z=61$, have no stable isotopes. It is notable that these are both examples with odd Z, but otherwise, the fact that this happens for these two elements, and not for any others, appears to be an 'accident'.

In the simple picture of β decay, conservation of energy and momentum would predict that almost all the available energy (corresponding to the energy difference between the initial and final

*However not necessarily stable to other forms of decay; for example, there are no stable nuclides with mass numbers 5 or 8. This fact—which seems to be a reflection of the extremely high stability of the α particle—is important in theories of the origin of the elements described in Chapter 3.

nuclides) should appear as kinetic energy of the emitted electron. This is not actually the case. Electrons are found to be emitted with a range of energies, from zero up to the maximum available. This observation (and other considerations, such as the conservation of angular momentum) led physicists to postulate the existence of another uncharged particle, the **neutrino**. According to this idea, the neutrino carries the 'missing' energy of the β decay. Neutrinos were detected directly in 1956, thus confirming the suggestion. All β decay and EC processes are accompanied by neutrino emission, although the neutrinos are not normally seen.

The half-lives of β decaying nuclei range from fractions of a second to values in excess of 10^{15} years. There is a general correlation with the energy release in the process; thus, nuclei a long way from the line of stability, where the energy curve is steep, tend to have short half-lives. Generally speaking, however, β decay is a very *slow* process on the time-scale expected for nuclear phenomena. It is in fact a manifestation of the **weak interaction** between elementary particles. Like the strong interaction responsible for nuclear stability, this is a force which acts only over very short distances, but as its name suggests, it is many orders of magnitude weaker in its effect. Neutrinos interact with matter *only* by the weak interaction, which is why they are so difficult to detect. Modern theories of particle physics are beginning to show how the weak and strong forces, together with the more familiar long-range electromagnetic force, can be explained as different manifestations of the same fundamental interaction. According to these theories, the very different ranges and strengths of the interactions are features of the 'low' energies at which we normally observe them. At extremely high energies, much larger than those involved in radioactive decay processes, the different forces should behave in a similar way, and under these conditions the 'weak' interaction can appear to be very strong. This idea is very important in some modern views on the origin of the universe, and it is also significant in **supernova** explosions of stars. As we shall see in the next chapter, such explosions form an essential route by which elements synthesized in stars are expelled into space. At the higher energies involved in supernovae, neutrinos can apparently interact quite strongly with matter, and are thought to play an important role in transporting energy from the centre of the star into the outer layers.

Nuclear reactions

The spontaneous decay processes of unstable nuclei are analogous to the chemists' **unimolecular reactions**, in which unstable molecules split up.

The synthesis of heavy elements from lighter ones requires the equivalent of **bimolecular** processes, resulting from the reactive collision of two species. Such reactions are not only important in understanding the origin of the naturally occurring elements, but as we shall see in the next section, they can also be used to generate artificial elements, unknown in nature because of their short half-lives.

One type of reaction is the simple **fusion** process, in which two nuclei combine to form a heavier one, the excess binding energy released in the form of a high-energy photon (γ). Examples of fusion reactions are:

$$^{1}H + {}^{2}H = {}^{3}He + \gamma \tag{2.7}$$

$$^{12}C + {}^{1}H = {}^{13}N + \gamma \tag{2.8}$$

$$^{12}C + {}^{4}He = {}^{16}O + \gamma. \tag{2.9}$$

Such 'simple' reactions are, in fact, rather uncommon among processes that can be studied in the laboratory. More frequent are reactions where the colliding particles **exchange** a number of nucleons. Two examples of exchange reactions are:

$$^{14}N + {}^{4}He = {}^{17}O + {}^{1}H \tag{2.10}$$

$$^{9}Be + {}^{4}He = {}^{12}C + {}^{1}n. \tag{2.11}$$

Both of these have a considerable historical interest. Reaction 2.10 was discovered by Rutherford in 1919 and was the first example of an artificially produced nuclear transformation. Reaction 2.11 was first performed by Chadwick in 1932 and led to the discovery of the neutron. Reaction 2.11 is still used sometimes as a small laboratory source of neutrons, by mixing beryllium with a natural α emitter such as radium.

Nuclear reactions are often presented in a notation different from the 'chemical' form given below. Reactions 2.10 and 2.11 can be written:

$$^{14}N(\alpha,p)^{17}O \tag{2.10'}$$

$$^{9}Be(\alpha,n)^{12}C. \tag{2.11'}$$

In this notation, the heavier initial and final nuclides are shown outside the brackets. The light bombarding particle is written inside the brackets, followed by the 'emitted' particle. Thus, (α,p) and (α,n) may be used generally to represent reactions in which a nucleus is transformed under ^{4}He bombardment, liberating either a proton or a neutron in the process. The fusion reactions discussed above may also be written in this form, using (p,γ) or (α,γ) to show that energy is liberated in the form of a γ photon. For example, reaction 2.8 is

$$^{12}C(p,\gamma)^{13}N. \tag{2.8'}$$

As explained in a previous section, nuclear collisions are subject to the same type of **Coulomb barrier** which inhibits α decay. Quantum mechanical tunnelling through the barrier can occur, but as with α decay, the rate of this is very small at low particle energies. For this reason, nuclear reactions studied in the laboratory generally require collision energies of at least several hundred keV. In the first reactions studied (for example, 2.10 and 2.11 above), this was provided by the kinetic energy of a naturally emitted α particle. The 'modern' era of nuclear physics was initiated by Cockcroft and Walton in 1932, when they observed the first nuclear reaction with protons, accelerated in the laboratory to energies of a few hundred keV. Contemporary particle accelerators are capable of producing beams of nuclei with energies up into the GeV range.

In the processes of nuclear synthesis in stars the energy of colliding nuclei is provided by the thermal distribution of velocities. Temperatures in excess of 10^7 K are required for these reactions, but even so, the kinetic energies involved are much lower than those normally necessary for studying nuclear reactions in the laboratory. In order to calculate the rates of such stellar processes, therefore, the rates measured in the laboratory have to be extrapolated to lower collision energies. This is normally done by using the theory of tunnelling through the barrier. For example, equations similar to 2.6 are used to predict how the tunnelling rates change with energy.

Although the synthetic reactions in stars take place at rather low energies, much higher energy collisions can occur in other situations, for example, in cosmic rays. Laboratory studies show that **spallation** is an important reaction under these conditions. In this process the kinetic energy of a bombarding particle (usually a proton) is sufficient to knock several small fragments out of a heavier nucleus. A typical spallation reaction may leave a nucleus with a mass number of 20 or so below that of the 'target', the remaining mass being distributed as a range of protons, neutrons, α particles and so on. An example is:

$$^{56}\text{Fe} + {}^{1}\text{H} = {}^{36}\text{Cl} + {}^{3}\text{H} + {}^{3}\text{He} + 2\,{}^{4}\text{He} + 3\,{}^{1}\text{H} + 4\text{n}. \qquad (2.12)$$

Spallation reactions in cosmic rays are thought to be the major source of the light elements Li, Be, and B (see Chapter 3). Similar reactions performed in particle accelerators are sometimes used as an alternative to nuclear fission reactors (see below) to produce neutrons for research purposes.

Because they have no charge, neutrons do not experience the Coulomb barrier which is important in other nuclear reactions. For this reason, they can produce reactions at very much lower kinetic energies.

A common result of neutron bombardment is the (n,γ) process of **neutron capture**; for example,

$$^{98}\text{Mo} + \text{n} = {}^{99}\text{Mo} + \gamma. \tag{2.13}$$

As in this example, neutron capture by a stable nuclide produces one on the neutron-rich side of line of stability, which may be unstable to β decay. For example, ^{99}Mo undergoes β^- decay with a half-life of 67 hours, to form ^{99}Tc—in this case, an isotope of an element that does not occur naturally on Earth. Neutron capture is the most important process by which elements heavier than iron are produced in stars. At low energies, neutron capture may be studied by exposing samples to the neutron flux present in a nuclear reactor. It is often used, as in the above example to produce radioactive nuclides for research purposes.

Just as elements differ widely in their chemical reactivity, so they also show a large variation in their tendency to undergo nuclear reactions. Generally speaking, the more 'stable' nuclei, with even values of Z and N, and especially when one of these is a magic number, are less reactive. The greater abundance of such stable nuclei in the universe is controlled to a large extent by this trend. Under conditions where nuclear reactions are occurring, the more reactive species will be consumed rapidly, and so will not build up in large concentrations. The stable nuclei, on the other hand, will tend to accumulate, as the reactions which destroy them will be relatively slow.

A good example of the trends in nuclear reactivity is shown in the relative probability of neutron capture. Figure 2.12 shows a plot of **cross-sections** for the capture of 25 keV neutrons. Such cross-sections represent the 'effective area' presented by the target nuclei to incident neutrons, and determine the relative rate of reaction of different nuclides. Both the odd–even alternation and magic number effects are shown clearly in this diagram, with the more stable nuclei having lower cross-sections. The large drops in cross-section close to magic-number configurations (with the elements concerned shown on the plot) are especially notable. In the next chapter we shall show how the trends shown in Fig. 2.12 are important in understanding the relative abundances of the heavy elements.

It was mentioned earlier that spontaneous fission into two nearly equal fragments is a possible mode of decay of some very heavy nuclei. In 1938 Meitner and Frisch discovered that fission can be induced in ^{235}U by the absorption of a neutron. They demonstrated this by chemically identifying barium as one of the products of the neutron bombardment of uranium. It was soon realized that more neutrons are also released in the fission process. This is inevitable when a heavy element splits into lighter

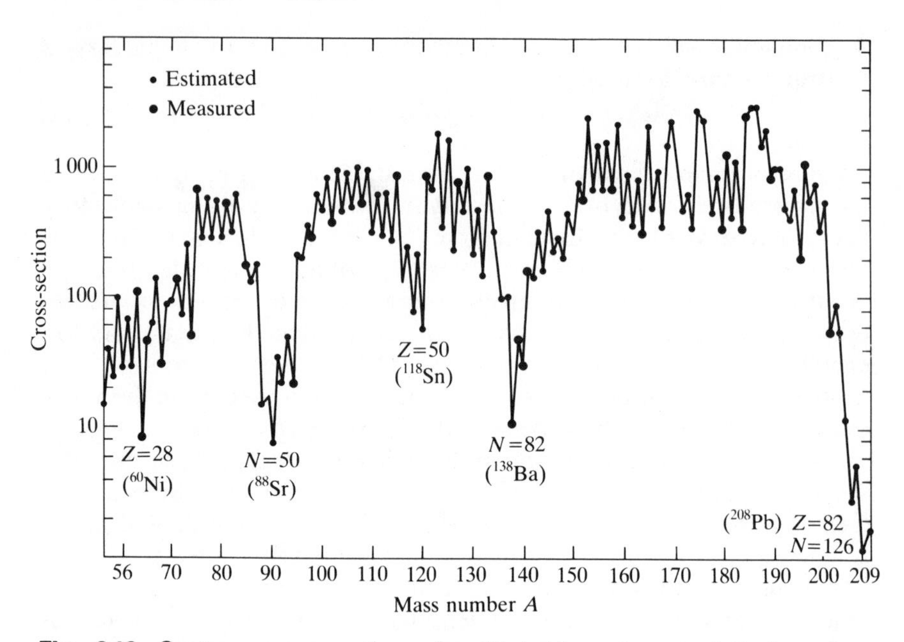

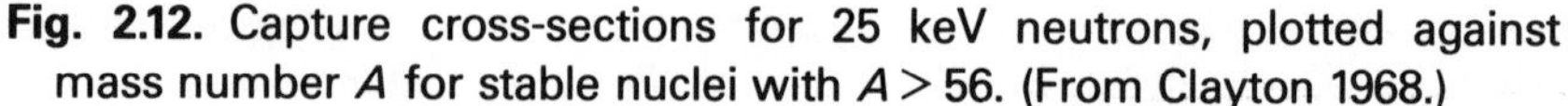

Fig. 2.12. Capture cross-sections for 25 keV neutrons, plotted against mass number A for stable nuclei with $A > 56$. (From Clayton 1968.)

ones, since as explained previously heavy nuclei contain a greater proportion of neutrons than the lighter products. Therefore, a **chain reaction** can be established, with the neutrons produced by one fission event initiating fission of further nuclei. The energy released is about 75 MeV per starting nucleus, and the enormous value of this compared with the energies of chemical reactions (a few eV per molecule), leads to the application of chain reactions in nuclear weapons and in thermal reactors for nuclear power production. The exothermic fusion reaction of light elements, especially:

$$^{2}\text{H} + {}^{3}\text{H} = {}^{4}\text{He} + \text{n} \tag{2.14}$$

gives even more energy output per mass of starting material. This reaction is used in thermonuclear weapons, and is being investigated as a possible source of future energy, although many technical problems remain to be overcome.

Radioactive elements

The nuclear reactions occurring in stars produce not only stable nuclides, but also unstable, radioactive ones. Most of these decay quite rapidly on an astronomical or geological time scale, although in a few cases the products of this decay can be detected and, as we shall see in later chapters, give very interesting information about the history of the solar system. However, there are several radioactive nuclides with half-lives sufficiently long that appreciable amounts remain on the Earth today. The most important of these are listed in Table 2.1. A more complete listing of the naturally occurring isotopes of the elements is given in Appendix B, and shows other long-lived radioactive nuclides, many with half-lives greater than 10^{11} years. It is notable that all radioactive nuclides present on Earth in significant amounts have half-lives longer than about 10^9 years. This value gives an order-of-magnitude estimate of the age of the Earth, if it is assumed that shorter-lived species were originally present, but have long since decayed. As explained in Chapter 6, more precise dating methods depend on measuring the amounts of radioactive elements remaining in rocks, together with the amount of the **daughter** species formed by decay.

Table 2.1
Some long-lived radioactive nuclides

Nuclide	*Decay mode*	*Stable products*	*Half-life (years)*
^{40}K	β, EC	^{40}Ca, ^{40}Ar	1.3×10^9
^{87}Rb	β^-	^{87}Sr	4.9×10^{10}
^{147}Sm	α	^{143}Nd	1.1×10^{11}
^{187}Re	β^-	^{187}Os	4×10^{10}
^{232}Th	α ... (series)	^{208}Pb	1.4×10^{10}
^{235}U	α ... (series)	^{207}Pb	7.0×10^8
^{238}U	α ... (series)	^{206}Pb	4.5×10^9

The existence of natural radioactive elements is important for several reasons. They contribute (along with cosmic rays) to the background radiation to which all living organisms, including ourselves, are exposed. This radiation can cause genetic mutations and may, therefore, have played an important role in evolution. In global terms, the energy liberated within the Earth by the radioactive decay of Th, U, and ^{40}K is quite significant, and provides the solution to a problem which confronted nineteenth century geologists. Physicists such as Lord Kelvin calculated the rate of heat loss from the Earth's interior, and showed that it would be expected to cool down on a time-scale of a few million years. Even at that time, it was realised that the Earth must be much older (its age is now reckoned at about 4.6 billion years), but no source of energy was then known which could keep it hot. It is, indeed, radioactive decay which provides this source; its importance to our present story is that it fuels many of the geological processes which have led to the present distribution of elements in the Earth's crust. Radioactive decay also influences the abundance of some elements on the Earth in a much more direct way. For example, most argon in the atmosphere is a product of the β decay of ^{40}K, and helium results from the α decay of heavy elements such as uranium. As mentioned above, the presence of radioactive decay products, which have been formed at rates that can be calculated from the known half-lives of the parent species, offers a unique opportunity for geological dating, and it is on this basis that we know, for example, the age of the Earth.

In addition to long-lived isotopes, there are very small amounts of several much shorter-lived ones on the Earth. Some of these have been produced artificially in the last 50 years, but there are also some naturally-occurring nuclear processes which can generate them. Small quantities of short-lived species are produced in the upper atmosphere, as a result of nuclear reactions induced by cosmic rays. The best known of these is ^{14}C, with a half-life of 5730 years. The decay of uranium and thorium also gives rise to short-lived radioactive elements, as is illustrated by the ^{238}U **decay series** shown in Fig. 2.13. Other series are formed from the decay of the rarer isotope ^{235}U and from ^{232}Th. The decay products of these nuclides are themselves radioactive, and so decay further until a stable isotope of lead is reached. The nuclides of the series in Fig. 2.13 are plotted against their atomic number (Z) and their neutron number (N). The characteristic changes involved in α and β decay process are shown by arrows in different directions. Thus, starting from a 'parent' nuclide (Z, N), α decay leads to a 'daughter' with ($Z-2$, $N-2$), and β^- decay to one with ($Z+1$, $N-1$). In addition to the kinetic energy carried by the α particle or the electron in these decay processes, some energy also appears as photons, that is γ rays.

A sample of natural uranium will obviously contain isotopes of all the elements forming part of the series shown in Fig. 2.13. It can be shown that the concentration of each radioactive element in the series will build up to a value proportional to its own half-life. The decay products can in principle be separated and identified by chemical means, and the longest-lived members of the ^{238}U series, **radium** and **polonium**, were first discovered in this way by Marie Curie in 1898. **Radon** (a noble gas), **actinium**, and **protoactinium** were identified later. The two remaining elements between bismuth and uranium, **astatine** and **francium**, have very short half-lives, however, and occur naturally in exceedingly minute amounts. Detectable quantities of these elements were first produced by artificial means.

From the chemist's point of view, one of the most interesting results from the study of nuclear reactions has been the production of

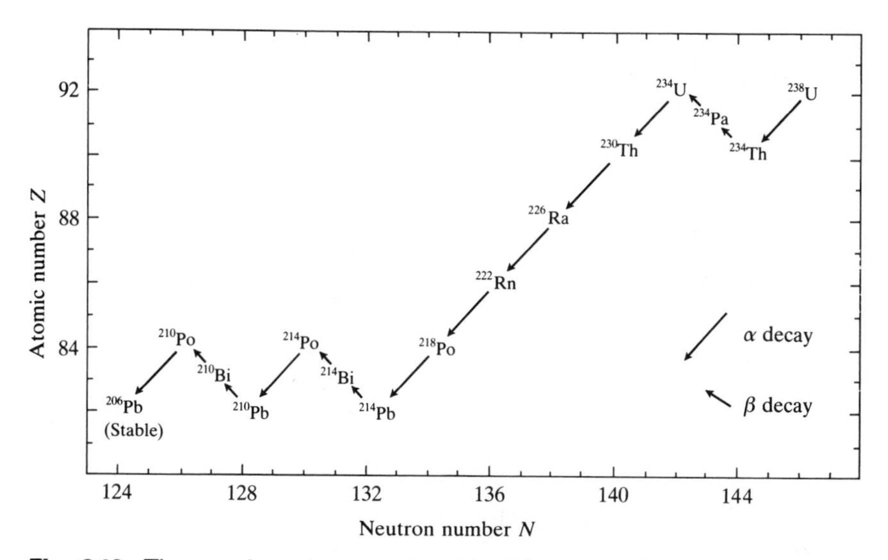

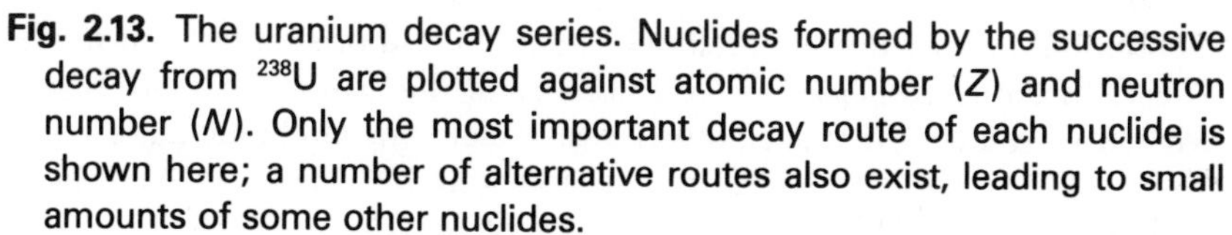

Fig. 2.13. The uranium decay series. Nuclides formed by the successive decay from ^{238}U are plotted against atomic number (Z) and neutron number (N). Only the most important decay route of each nuclide is shown here; a number of alternative routes also exist, leading to small amounts of some other nuclides.

previously unknown elements. It was shown previously how neutron bombardment of ^{98}Mo can lead to the production of **technetium** ($Z = 43$). This element and the missing lanthanide element **promethium** ($Z = 57$) were first observed as fission products from ^{235}U in nuclear reactors. Since 1940, largely through the work of Seaborg and his collaborators, several elements beyond the heaviest naturally occurring one (uranium) have also been discovered. Some of these can also be made in nuclear reactors. For example, **neptunium** ($Z = 93$) and **plutonium** ($Z = 94$) are generated following neutron capture by ^{238}U in the sequence:

$$^{238}U(n,\gamma)^{239}U(\beta^-)^{239}Np(\beta^-)^{239}Pu. \tag{2.14}$$

^{239}Pu is manufactured in multi-kilogram quantities, since, like ^{235}U, it undergoes neutron-induced fission and is used in nuclear weapons. A few of the succeeding elements can also be produced by successive neutron capture (essentially the same as the *r*-process described in the next chapter), and were first identified in fall-out following the atmospheric testing of nuclear weapons. The synthesis of heavier transuranium elements, however, requires energetically unfavourable nuclear fusion reactions, using ions of very high kinetic energy produced in an accelerator. Thus, element 102, **nobelium**, was first made by bombarding ^{246}Cm with ^{12}C, and a more recent example is the discovery of element 108 using ^{58}Fe ions accelerated to an energy of 292 MeV, to produce the reaction:

$$^{58}Fe + ^{208}Pb \rightarrow ^{265}108 + n \tag{2.15}$$

The artificial synthesis of transuranium elements has made it possible to verify the hypothesis first made by Seaborg, that the elements between $Z=90$ (thorium) and $Z=103$ (lawrencium) form an **actinide** series associated with the filling of the $5f$ shell and, therefore, analogous to the lanthanides. Nevertheless, the intense radioactivity of these elements and the increasingly short half-lives of the later ones make chemical investigations very difficult. For those most recently discovered (at the time of writing there are unconfirmed reports of element 110), the half-lives are measured in milliseconds. Thus, it might seem unlikely that our study of the chemical properties of elements can be extended much beyond the actinide series. However, calculations based on the nuclear shell model predict that there should be an 'island of stability' for nuclei with Z in the range 114–116, and with N in the range 178–180. It was originally predicted that such **super-heavy** elements might have life-times comparable to the age of the Earth, and searches for them have been made in minerals. Other attempts to detect them have concentrated on cosmic rays. A few positive claims have been made, but these have been disputed and at present there seems to be no good evidence for the natural occurrence of super-heavy elements. More recent calculations suggest that half-lives might be in the region of a thousand years, which provides a strong motivation for the continuing attempts to synthesize them using accelerated ions. For example, a possible route to the production of $^{296}116$ would be to bombard ^{248}Cm with ^{48}Ca. Experiments along these lines have so far given disappointingly negative results. Nevertheless, the theorists are still hopeful, as the heaviest elements so far discovered seem to undergo spontaneous fission much more slowly than would be expected from simple extrapolation from lighter ones. Thus, there are still grounds for believing that the predictions about super-heavy elements are correct, and that they will fairly soon be discovered.

Summary

Chemical forces are provided by the electrostatic attraction of electrons to nuclei. In nuclei themselves, the electrostatic interaction between protons is repulsive, and is counteracted by the attractive **strong interaction**, a short-range force acting equally between protons and neutron. The balance between the two forces results in ^{56}Fe being the most stable nucleus, so that it is energetically favourable for lighter nuclei to fuse together and for heavier ones to break up, for example by α decay. However, the same balance of forces gives large energy barriers for such reactions, so that fusion can only take place if colliding nuclei have high kinetic energies, and α decay occurs extremely slowly for elements up to bismuth.

The chemical regularities displayed in the periodic table can be explained by the quantum theory, which predicts the shell structure of atoms. **Electropositive** elements on the left-hand side of the table tend to lose electrons in compound formation while **electronegative** elements on the right gain electrons. Thus, compounds between very electropositive and very electronegative electrons are described by the ionic model. On the other hand, elements of intermediate electronegativity normally form less ionic compounds, in which electron sharing or covalent bonding is important.

Application of the shell model to nuclei explains the existence of **magic numbers** of protons or neutrons which give nuclei extra stability. There is also a **pairing force** which makes nuclei with even numbers of protons and/or neutrons more stable than those with odd numbers. These trends are both reflected in the reactivity of nuclei and, thus, influence the overall abundance of elements. An additional force, the **weak interaction**, gives rise to β decay processes which change protons into neutrons or vice versa, and this, taken with the other factors controlling nuclear stability, limits the numbers of stable isotopes of each element.

A number of long- and short-lived radioactive elements occur naturally on the Earth. In the last 50 years this list has been augmented by the use of nuclear reactions to synthesize artificial elements, especially ones heavier than uranium.

Further reading

Non-technical accounts of the fundamental forces of nature are given by Davies (1986), and of quantum mechanics by Gribbin (1985).

Chemical properties in relation to atomic structure and the periodic table are treated in numerous books on inorganic chemistry. Puddephatt and Monaghan (1986) give a brief account; among longer books those by

Huheey (1983) and especially Sanderson (1967) concentrate more on systematics and periodic trends; the ones by Cotton and Wilkinson (1988), and Greenwood and Earnshaw (1985) give detailed, element-by-element, descriptions. Fergusson (1982) presents an account of the chemistry of the elements in relation to their occurrence on the Earth.

Pearson (1986) gives a reasonably straightforward account of nuclear properties. More detailed discussions of the basic physics can be found in Burcham (1979), and of applications in Friedlander *et al.* (1981). There is a review by Seaborg and Loveland (1987) of attempts to make super-heavy elements.

3
The origin of the elements

What is possible in the Cavendish Laboratory may not be too difficult in the sun.
A. Eddington (1920)

Where did the elements come from? In principle, they could have been present at the beginning of the universe, or have been formed somehow in its very early stages. Alternatively, they may have been made later, in processes which might be continuing today. In fact, there are very strong reasons—both theoretical and observational—for believing that most elements were not present in the early universe. According to the widely accepted hot big bang theory, the universe started in a state of such high temperature that no nuclei could be stable. Although the very earliest stages are still a matter of dispute, it is clear that protons and electrons—the constituents of hydrogen—were the first stable constituents of ordinary matter to be formed. Early versions of the big bang theory suggested that all the elements might be formed very soon, but this is now known to be impossible, for reasons that will be explained below. The spectra of very old stars, formed over 10 billion years ago, do indeed show a marked deficiency of all elements except hydrogen and helium. Only these elements were formed in large amounts in the early universe; the others have been synthesized since.

Since the discovery of nuclear reactions, it has been realized that only such processes can account for the prodigious energy output of stars. As explained in Chapter 2, the fusion of lighter elements to form heavier ones, at least up to iron, is a strongly exothermic process, and gives energies many orders of magnitude larger than can be obtained from chemical reactions. The idea that *all* the elements could be synthesized in stars does not follow automatically. One of the difficulties which any theory of the origin of the elements must face is that their observed abundances cannot be explained by any single type of process, however unusual the conditions might be. It seems that elements in different mass ranges must be produced in very different environments. Most modern ideas about the synthesis of the elements from the work of Bethe in the

1930s, but it was only with the classic paper of Burbidge, Burbidge, Fowler, and Hoyle, published in 1957, that the routes leading to the production of most elements in stars were established. It is now generally accepted that the vast majority of elements are made in this way. Although there are still a few uncertainties about some of the *astronomical* details (because this is not a process that is very accessible to normal observational methods) the basic nuclear reactions seem to be well understood, and will be explained in this chapter.

The early universe

The modern theory of the origin of the universe stems from observations that atomic lines measured in the spectra of distant galaxies have longer wavelengths than those found on Earth. Although other explanations for this **red-shift** have been suggested, nearly all astronomers believe that the straightforward interpretation is correct, and that distant galaxies are receding from us. The red-shift is thus often explained as the **Doppler effect** in light emitted by moving objects, although it is better to think of it as a consequence of the general 'stretching' of light waves by the expansion of space itself. According to Hubble's law the apparent velocity of recession (v) is proportional to the distance (d) of a galaxy from our own:

$$v = Hd \tag{3.1}$$

where H is **Hubble's constant**. The most natural interpretation of Hubble's law is that the universe is in a state of uniform expansion, so that the distances between all galaxies are increasing. Extrapolating back into the past, it appears the density of matter must have been greater at earlier times than it is now. Continuing this extrapolation, one would arrive at a moment in the distant past when the density was apparently **infinite**. This conclusion is remarkable, and other interpretations of the expansion of the universe has been put forward. One such is the **steady state theory**, in which it is assumed that the expansion is compensated by the **continuous creation** of matter in space, so that the average density of the universe does not change with time. The steady state theory avoids the need for the universe to have a definite beginning. There is now strong evidence against this idea, however, and in favour of the view that the universe *is* evolving, and did indeed have a definite beginning between 10 and 20 billion years ago.*

*If the universe has expanded at a constant rate since its origin, its age must be the inverse of the Hubble constant, $1/H$. There are two problems with this estimate. In the first place, it is difficult to determine H accurately, because of uncertainties in the distance of galaxies very far from our own. Secondly, the rate of expansion may be changing. The

According to the currently accepted **hot big bang** theory, the universe began in a state of unimaginably high density and temperature. As it expanded, it cooled. One second after the beginning, the temperature was about 10^{10} K, and before this point, no elements except hydrogen could exist, because the average kinetic energy of particles was so high that any other nuclei formed would immediately dissociate. In fact, at even higher temperatures in the first millisecond or so, the energies were so large that most of the matter was present in the form of exotic elementary particles that are unstable under normal conditions. The high energies involved in the earliest moments are quite inaccessible to laboratory study and so an understanding of these stages has come about only through the development of better theories of elementary particles.

One of the most puzzling things about the universe is why it has any matter in it at all. This difficulty arises because most elementary particles, such as protons, neutrons, and electrons, have corresponding antiparticles. These can be produced on Earth in high energy accelerators, but are annihilated when they meet normal matter, the energy (equal to mc^2 per particle, according to equation 2.4 on p. 29) being liberated, for example in the form of photons. Thus, an antiproton $\bar{p}$ will annihilate a proton (p):

$$\bar{p} + p = 2\gamma \quad \text{(or other particle/antiparticle pairs).} \qquad (3.2)$$

At the enormously high temperatures of the big bang, large amounts of antimatter were present. On cooling below some 10^{12} K, the equilibrium in equation 3.2 shifted to the right, and all the antiprotons disappeared. Since some normal protons remained, however, there must have been a slight excess of these to start with. In fact, the excess of matter over antimatter was exceedingly minute, as in the universe today there is only one proton per 10^9 or more photons. It is this very small, but non-zero excess that is puzzling. One possibility is that both matter and antimatter were present in equal amounts, and survived by somehow becoming separated. If this is correct, there must be distant galaxies composed of antimatter. It is thought to be unlikely, however, partly because it is very difficult to understand how the separation might have occurred. According to current theories, excess protons were created in the first fraction of a second by a so-called **hyperweak force**. One interesting prediction is that the hyperweak force which created matter will also eventually destroy it again; thus, protons—and indeed, all nuclei—are

most commonly accepted model of the universe suggests that the expansion is slowing down due to gravitational forces, so that $1/H$ gives an *overestimate* of the age of the universe. The best current estimates are that $1/H$ is nearly 20 billion years, and that the big bang happened about 16 billion years ago. It will be shown in Chapter 6 that information on elemental abundances can be used to obtain an independent estimate of the age of the universe.

unstable, and should decay with a half-life somewhere around 10^{32} years. Experiments are in progress to test this.

Although the very early stages are still uncertain, the picture becomes much clearer after the point where matter as we know it could form. After 1 second, the universe was composed of more familiar particles—protons, neutrons, electrons, neutrinos, and photons. Free neutrons would still be present in equilibrium at these temperatures, although they would soon start to undergo β^- decay into protons, electrons, and antineutrinos:

$$n = p^+ + e^- + \bar{\upsilon}. \tag{3.3}$$

As the temperature fell to around 10^9 K, more complex nuclei could start building up, by processes of fusion and neutron capture. The most important reactions, which were completed in a few minutes, were:

$$p + n = {}^2H + \gamma \tag{3.4}$$

$${}^2H + {}^2H = {}^3H + p \tag{3.5}$$

$${}^2H + {}^2H = {}^3He + n \tag{3.6}$$

$${}^3He + n = {}^4He + \gamma \tag{3.7}$$

$${}^3H + p = {}^4He + \gamma \tag{3.8}$$

Under the conditions present, it was certainly the first of these reactions—the formation of deuterons (^{2}H)—that constituted the rate-limiting step. This is because of the rather low binding energy of deuterons (2.2 MeV), so that they would be dissociated almost as rapidly as they formed. The remaining reactions occurred rapidly, and ^{4}He was by far the major product. The proportions of deuterium and ^{3}He remaining from this early stage of nucleosynthesis are very sensitive functions of the density of matter and so are a useful indicator of the conditions present at that time. On the other hand, the amount of helium produced depends less strongly on the density, and is largely determined by the equilibrium ratio of neutrons to protons at the temperature when the nuclear reactions can begin. Calculations predict a He/H atomic ratio of nearly 1:10, or about 25 per cent helium by weight. This is the abundance observed in stars, and is one of the strongest pieces of evidence that the big bang theory is correct.

It was originally thought that all the elements heavier than helium might also be produced in the big bang. It is now recognized that this is impossible, for a number of reasons. By the time the reactions 3.4–3.8 can occur, the temperature and density are falling rapidly. The fusion of nuclei heavier than hydrogen involves a correspondingly larger Coulomb barrier, and requires the sustained conditions of high temperature and density present in stars. Neutron capture might present an alternative route avoiding the Coulomb barrier. However, apart from the fact that free neutrons are decaying rapidly in this phase, such a process is blocked by the non-existence of stable nuclei with mass numbers 5 and 8. These form gaps which are unbridgeable in any simple process of building up one mass number at a time. In fact, the gaps *can* be bridged, by reactions such as:

$$^{4}He + ^{3}H = ^{7}Li \quad (3.9)$$

However, because of the falling temperature and density, only very minute amounts of nuclides such as ^{7}Li can be made. Most of the ^{7}Li present today may have come from reaction 3.9, but it is quite impossible that appreciable amounts of any heavier elements can have been produced in the early universe.

In addition to the matter formed—largely left as hydrogen and helium—the big bang gave rise to large amounts of electromagnetic radiation in space. At the time helium was synthesized, of course, the temperature was far too high for neutral atoms to be present, and the matter would have been in the form of nuclei and free electrons. Such charged particles form a **plasma** which interacts very strongly with all wavelengths of electromagnetic radiation, and would thus keep the radiation in thermal equilibrium with the matter. This gives rise to what is known as the **black-body distribution** of radiation, illustrated for different temperatures in Fig. 3.1. As the universe cooled, a point was reached, about a million years after the beginning, when the temperature was low enough (a few thousand Kelvin) for electrons and nuclei to combine to form neutral atoms. Such atoms do not interact strongly with photons, so that matter and radiation became 'uncoupled'. They could then evolve independently. The matter—or at least a fair proportion of it—condensed under gravitational forces to form stars. The radiation remained in space and, as the universe expanded, the electromagnetic waves were 'stretched' or red-shifted in the same way as light from distant galaxies. At the time of uncoupling, the black-body radiation was concentrated in the visible region of the spectrum (see Fig. 3.1). However, the universe has expanded by a factor of several thousand since this time, and the radiation is now largely in the millimetre region of the spectrum, corresponding to an effective temperature of 2.7 K. This background radiation in intergalactic space is a kind of 'cosmic echo' of the big bang, and its observation, first reported by Penzias and Wilson in 1965, is another crucial test of the theory.

As our discussion has implied, current ideas about the origin and structure of the universe are changing rapidly. Apart from the problems of the origin of elementary particles, there is also some difficulty in

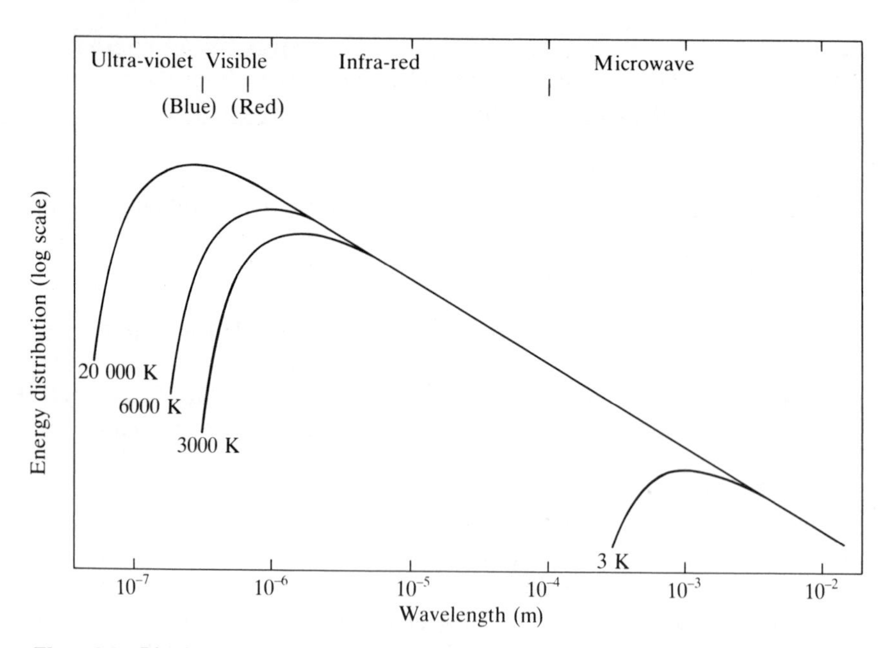

Fig. 3.1. Black-body radiation curves, showing the wave-length distribution of photons in thermal equilibrium at different temperatures, on a logarithmic scale. The different regions of the electromagnetic spectrum are marked.

understanding the amount of matter in the universe, and its apparently very uneven distribution in clusters and 'superclusters' of galaxies. It seems that there may be much more matter than appears to us. Indeed, some cosmologists now believe that the universe may be full of invisible particles which are only apparent through their gravitational attraction to ordinary matter. These have been called **weakly interacting massive particles** or WIMPs. Whether such ideas are correct remains to be seen. However, the WIMP theory does offer a possible solution to another famous problem in astrophysics: this is the case of the 'missing solar neutrinos', discussed later in this chapter.

The evolution of stars

Some properties of a star—its surface temperature, luminosity, size, and sometimes its mass—can be deduced from purely astronomical observations. The application of basic laws of physics then allows many details of the interior, such as the central temperature, to be estimated even without understanding the energy source involved. For this reason many features of the structure and evolution of stars were known to astronomers before the nuclear reactions which fuel them were understood.

We saw in Chapter 1 how the elemental composition of a star can be studied from absorption or emission lines present in its spectrum. The spectrum can also show the **surface temperature** of a star. This information comes partly from the overall wavelength distribution of radiation, and partly from a knowledge of the particular species—ions, atoms, and in cool stars even some simple molecules—revealed by the spectrum. Figure 3.1 showed the distribution of radiation emitted by 'black bodies' in thermal equilibrium at different temperatures. The radiation from the Sun, when corrected for absorption by the Earth's atmosphere, corresponds closely to that from a black body at a temperature around 6000 K. Similar measurements on other stars show surface temperatures ranging from 3000 K to over 20 000 K. These temperatures are reflected in the apparent colours of the stars, as can be seen from Fig. 3.1. For example, radiation from the Sun is fairly evenly distributed throughout the visible region of the spectrum, but with some fall-off towards the blue end, giving a yellowish colour. Stars cooler than the Sun (with surface temperatures around 3000–4000 K) will appear reddish, and hotter ones may look white (around 10 000 K) or even blue (20 000 K). This range of colours can be seen by looking at stars in the night sky.* It should be emphasized that it is only the *surface* temperature which is directly revealed in this way; the temperatures at the centres of stars are much greater, ranging from 10^7 to above 10^9 K.

The other obvious way in which stars vary in external appearance is in their brightness. Of course, the apparent brightness of a star—which astronomers call its **magnitude**—depends partly on its distance from ourselves. For stars not too far away, distances can be found by the method of parallax, that is, from the extent to which their apparent positions against a background of much more distant stars and galaxies vary during the year, as the Earth moves in its orbit around the Sun. Only the nearest stars can be located in this way, and in other cases, astronomers are forced to rely on less direct methods. When the distance is known, an estimate can be obtained of the intrinsic brightness or **luminosity** of a star. The most important classification scheme for stars uses a plot which shows both their surface temperature and luminosity.

*The constellation of Orion, which dominates the night sky in the winter months, is particularly good for showing this. The brightest stars in Orion are the red Betelgeuse and the blue Rigel, shown on Fig. 3.2. Also of interest is the famous Orion Nebula, a large region of luminous gas where new stars—some probably with planetary systems—are being formed.

This is the so-called **Hertzsprung–Russell** (H–R) **diagram**, illustrated in Fig. 3.2.

From the position of a star on the H–R diagram, its radius can also be deduced. This is because the luminosity (L) of a 'black body' is related to its surface area ($4\pi R^2$) and its absolute temperature (T) by **Stefan's law**:

$$L = 4\pi R^2 \sigma T^4 \tag{3.10}$$

Cool stars with a high luminosity must therefore be very large and, conversely, hot dim stars must be small. It can be seen that a high

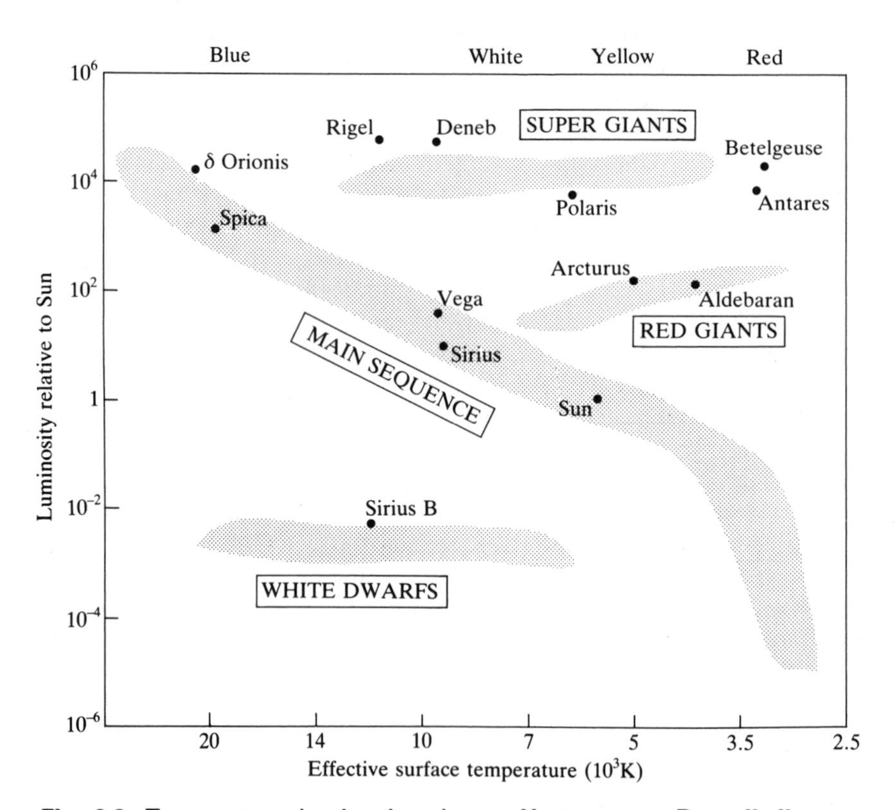

Fig. 3.2. Temperature–luminosity plot, or **Hertzsprung–Russell diagram,** of stars. A number of prominant stars are shown by name. The visible colours of stars are indicated at the top, and the different classes of stars labelled in the diagram. Many prominant stars are giants or super-giants, but these are atypical, and a majority of stars are found to lie on the Main Sequence.

proportion of stars is clustered along a diagonal line in the H–R diagram, called the **main sequence**. Most main sequence stars have sizes within a factor of ten of that of the Sun, increasing towards top end where the hotter, more luminous stars are situated. Stars in other regions of the diagram may be very different in size, and are given self-explanatory names such as **red giants**, **supergiants**, and **white dwarfs**.

Some of the brightest stars visible in the Northern Hemisphere are indicated in Fig. 3.2. The Sun itself is a main-sequence star of rather average brightness and temperature. A little further up on the main sequence is Sirius, the 'Dog Star', and higher still are some of the blue stars in Orion. Among the red giants are Aldebaran and Arcturus, the brightest stars in the constellations of Taurus and Boötes, respectively. The brightest stars in Orion are supergiants: the cool red Betelgeuse, and the hot blue Rigel. To give some idea of the size of these, the radius of Betelgeuse is so large that it would extend out to the orbit of Jupiter in our solar system.

The ways in which stars evolve on the H–R diagram can be deduced by looking at clusters of stars which were formed at about the same time. It appears that stars spend the major portion of their lifetime on the main sequence. As discussed in the next section, the principal energy source for main sequence stars is that of **hydrogen burning**, a series of nuclear reactions converting hydrogen into helium. A star does not evolve up or down the main sequence, but remains at a point which is determined by its mass: heavier stars are hotter and brighter. The masses of main sequence stars range from one-tenth of that of the Sun at the lowest part, to some 50 or 100 solar masses at the upper end. The range of luminosities shown on the diagram, however, is vastly greater, covering a factor of more than 10^{10}. It follows that the relative rate of energy production must be much greater in heavier stars, so that they burn up more quickly, and stay in their hydrogen burning phase on the main sequence for correspondingly shorter times. The Sun, for example, has been on the main sequence for around 4.5 billion years and will remain there for another 5 billion years or so. Very bright heavy stars may only stay there for a few million years. On the other hand, some very old clusters still contain dim stars, smaller than the Sun, that have been on the main sequence for more than 10 billion years, and will stay there for a long time to come.

The subsequent evolution of a star also depends on its mass. An 'average' star like the Sun will become a red giant when the hydrogen in its core is used up, and further nuclear processes such as **helium burning** begin. This is not expected to happen for another 5 billion years, which is fortunate as the Sun will expand so much that the inner planets (including the Earth) will be engulfed and vapourized. Following the red giant stage, the outer layers will be thrown off, and the sun will end up as a white dwarf—a dim star with very small radius and high density, which will eventually cool and fade from sight. Heavier stars evolve into supergiants, and it is the nuclear reactions in the interiors of these stars that gave rise to most of the heavy elements in the universe. Many supergiants end their life in colossal explosions—supernovae—which are not only important in the synthesis of many elements, but also play an essential role in ejecting elements out into space.

A star begins life with the contraction of a large mass of gas under gravitational forces. The potential energy released heats the gas, until the temperature at the centre is above 10^7 K, when nuclear reactions can begin. The energy produced by these reactions prevents further collapse, and stabilizes the size and luminosity of the star in its position on the main sequence. The heavier the star, the larger is the gravitational energy released in contraction. Therefore, higher central temperatures are reached and the nuclear reactions proceed faster. This is why heavier main sequence stars are so much more luminous and burn up more quickly.

Until recently, the very early stages in stellar evolution—the actual formation of main sequence stars—were much less understood than some of the later stages. Very young stars cannot be observed with optical telescopes, because they are generally found in regions surrounded by opaque clouds of gas and dust. Several regions of star formation—including the famous Orion nebula—are now known, as a result of observations with telescopes sensitive to infra-red radiation, which is less strongly scattered than visible light. It is believed that the energy output of very young stars may fluctuate quite markedly before they reach a stable state on the main sequence. A fair amount of matter may also be emitted in these early stages, producing a 'stellar wind' which eventually blows away much of the dust and gas originally surrounding the star. These features must have played an important part in the formation of the solar system, which will be discussed in the next Chapter.

It appears that the processes of star birth and death may be related in an intriguing way. Regions of our own and other galaxies where new stars are forming also show the remains of 'dead' stars, such as the remnants of supernovae explosions. The shock-waves emanating from supernova may act as a 'trigger' which initiates the collapse of a gas cloud to form new stars. There is considerable evidence, discussed in Chapter 6, that the birth of the Sun and the formation of the solar system started only a few million years after the explosion of a nearby star. Observations also show that this coupled process of star birth and death is particularly active in the spiral arms of galaxies; indeed, it may be one of the mechanisms which leads to the formation of spiral galaxies, although this process is still not well understood.

Hydrogen burning

The fusion of four hydrogen atoms to give helium liberates about 24 MeV per atom formed. This is by far the greatest energy release per mass of reactants in any simple nuclear process, and it provides the energy source for stars during the major, main-sequence phase of their active life. The reaction which starts at temperatures around 10^7 K is based on the initial fusion of two protons, and the most important sequence is known as the **PPI chain**:

$$^{1}H + {}^{1}H = {}^{2}H + e^{+} + v \tag{3.11}$$

$$^{2}H + {}^{1}H = {}^{3}He + \gamma \tag{3.12}$$

$$^{3}He + {}^{3}He = {}^{4}He + 2\,{}^{1}H \tag{3.13}$$

All these reactions, of course, are inhibited by the Coulomb barrier discussed in Chapter 2. Tunnelling through the barrier is possible, but even for this to take place at an appreciable rate requires high collision energies. This is why very high temperatures are required, when the nuclei have high thermal kinetic energies. The first step, however, reaction 3.11, also involves the transformation of a proton into a neutron by a β decay process liberating a positron (e^{+}) and a neutrino (v). Like all β decay reactions, this depends on the **weak interaction** between elementary particles, in contrast to the remaining steps which involve only the strong and electromagnetic interactions. Reaction 3.11 is therefore exceedingly *slow*, even at energies where the Coulomb barrier is unimportant. The slowness of the initial step is the main reason why stars stay so long in their hydrogen-burning phase on the main sequence. It is an interesting fact that if the strong interaction between nucleons were only very slightly stronger, two protons could bind together to form the ^{2}He nucleus. Reaction 3.11 could proceed in two separate steps:

$$^{1}H + {}^{1}H = {}^{2}He + \gamma \tag{3.11a}$$

$$^{2}He = {}^{2}H + e^{+} + v \tag{3.11b}$$

A significant concentration of ^{2}He could build up and the overall reaction would be much faster. The consequences would be dramatic, hydrogen burning in most stars being finished in less than a million years, instead of lasting for 10^{10} years as in the case of the Sun. Under these conditions, the universe would certainly have developed in a very different way, and stellar evolution would probably have finished too quickly for planetary systems to form, or for living organisms to arise. On the other hand, if the strong interaction were significantly *weaker* than it is, no elements beyond hydrogen would have been formed at all. Thus, we seem to owe our existence to an 'accident' of physics—unexplained by current theories—which makes the ^{2}He nucleus unstable, but only just so.

Another important consequence of the slowness of reaction 3.11 is that the proportions of ^{2}H and ^{3}He formed are exceedingly small. In fact, these nuclides—left in small quantities in the universe as a consequence of the big bang—are actually consumed, not manufactured, in stars.

A small proportion of the ^{3}He formed in equation 3.12 reacts by a different route, and gives rise to the **PPII** and **PPIII** chains. The PPII reactions are:

$$^{3}\mathrm{He} + {}^{4}\mathrm{He} = {}^{7}\mathrm{Be} + \gamma \tag{3.14}$$

$$^{7}\mathrm{Be} + \mathrm{e}^{-} = {}^{7}\mathrm{Li} + \upsilon \tag{3.15}$$

$$^{7}\mathrm{Li} + {}^{1}\mathrm{H} = 2\,{}^{4}\mathrm{He} \tag{3.16}$$

The PPIII chain results from an alternative reaction of ^{7}Be:

$$^{7}\mathrm{Be} + {}^{1}\mathrm{H} = {}^{8}\mathrm{B} + \gamma \tag{3.17}$$

$$^{8}\mathrm{B} = 2\,{}^{4}\mathrm{He} + \mathrm{e}^{+} + \upsilon \tag{3.18}$$

Although they only occur as a small fraction of the total PP process, these reactions are interesting for two reasons. In the first place, they lead to the consumption of some of the ^{7}Li which is produced in very small amounts in the early universe. Secondly, the neutrinos produced in reactions 3.15 and 3.18 should have sufficiently high energies to be detectable on Earth. Most of the energy produced by the nuclear reactions deep in the interior of stars ends up as electromagnetic radiation, but this reaches the surface only slowly, and is modified by countless events of scattering, absorption and re-emission, and so on, on its way out. In fact, it takes several million years for energy of this kind to reach the surface of the Sun. By contrast, neutrinos, which take part only in weak interactions and have no electromagnetic properties, pass through matter with very little interaction. Thus, the neutrinos produced deep in the Sun can reach the Earth directly. Their weak interaction makes them difficult to detect, of course, but neutrinos with sufficient energy have a small probability of producing inverse β decay reactions such as:

$$^{37}\mathrm{Cl} + \upsilon = {}^{37}\mathrm{Ar} + \mathrm{e}^{-} \tag{3.19}$$

An experiment to count solar neutrinos has been running since 1967, using a large tank full of perchloroethylene, C_2Cl_4, in a deep mine in the USA. The idea is to detect the ^{37}Ar produced in reaction 3.19, by its β decay back to ^{37}Cl. The number of neutrinos detected is much less than expected, which is worrying as it suggests that the conditions present in the interior of the Sun are not so well understood as one would like to think. The most straightforward interpretation of the discrepancy is that the central temperature is about 10 per cent less than that required by conventional theories. A lower temperature would reduce the rate of the PPII and PPIII reactions, and bring the rate of neutrino production into line with the observations. Various possible explanations of the 'solar neutrino problem' have been proposed over the years. An interesting recent one involves the possibility of 'WIMPs'—the 'invisible' particles also suggested as a solution to some of the problems of large-scale structure in the universe (see p. 68). It is suggested that the Sun may contain a large number of these particles, attracted by its gravitational field. WIMPs in the centre of the Sun might interact sufficiently with ordinary matter to carry away some of the energy generated by the PP reactions. This 'unobservable' energy loss from the Sun would thus alter the calculations from which its central temperature is deduced, and allow this temperature to be reduced by the required amount.

In addition to the PP chains, there are other reaction schemes that can convert hydrogen into helium. An important one is the **CNO cycle**, shown in Fig. 3.3. This diagram uses the notation for nuclear reactions explained in Chapter 2 (see p. 53). The cycle starts with a small catalytic amount of ^{12}C initially present in the star. (p,γ) represents fusion reactions where a proton is absorbed by a nucleus, with energy emitted as γ radiation. These are interspersed with β^+ decays, and the cycle ends with (p,α), a reaction producing ^4He and regenerating the ^{12}C. Thus, the overall reaction is that four protons make one helium nucleus, as with the PP chains. As Fig. 3.3 shows, side reactions occur which lead to further cycles.

The relative importance of the PP and CNO processes in hydrogen-burning stars depends on a number of factors. In the first place, the CNO cycle avoids the weak-interaction 'bottleneck' of the first step in the PP chain, and so is potentially much faster, even with only small amounts of ^{12}C present. On the other hand, the Coulomb barriers involved in adding protons to more highly charged nuclei such as ^{12}C are much greater than those in the PP chain. The result is that the CNO cycle requires a higher temperature to start and becomes faster than the PP process only at temperatures above about 2×10^7 K. This is hotter than the interior temperature of the Sun, where the majority of hydrogen burning occurs by the PP chains. For heavier main-sequence stars, however, the central temperatures are higher and so the faster CNO cycle takes over. It is the

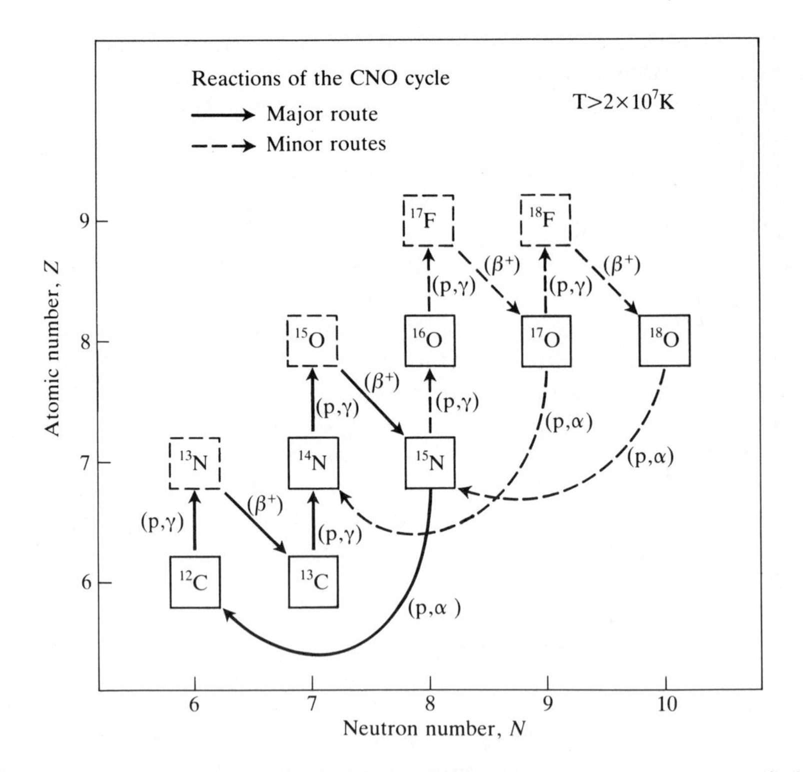

Fig. 3.3. The nuclear reactions of the CNO cycle, an important catalytic route for hydrogen burning in heavy stars. The notation is explained in Chapter 2 (see p. 53): (p,γ) represents a proton capture, and (p,α) a reaction which produces a ^{4}He nucleus by proton bombardment. The overall result of the cycle is to convert four hydrogens to one helium, with the initial ^{12}C being regenerated.

comparative speed of these reactions which leads to the much shorter lifetimes of heavy stars. Clearly, *some* elements heavier than helium must be present in the star to start with, if the CNO cycle can operate. It will be more important therefore, in the younger, so called **population I** stars in the spiral arms of our galaxy, than the older **population II** stars where the concentration of heavy elements is lower (see Chapter 1, p. 19). The first stars, formed very early on in the universe, presumably could not operate the CNO cycle.

As well as forming an important mechanism for hydrogen burning, the CNO cycle is probably the major source of the nuclides ^{13}C, ^{14}N, ^{15}N, and ^{17}O in the universe. The relative yield of each of these depends inversely on the rate of the reaction which consumes it in the CNO cycle.

Thus, ^{13}C and especially ^{15}N, which is one proton short of a magic number of both protons and neutrons, react quickly with another proton and are therefore present in lower steady-state concentrations than ^{14}N which reacts more slowly. It should be noted that carbon and oxygen have much larger overall abundances than nitrogen, because their commoner isotopes, ^{12}C and ^{16}O, are produced in the helium burning reactions discussed in the next section.

Helium burning and subsequent processes

As a result of the hydrogen burning reactions, a central core of helium builds up in the star. When this becomes very depleted in hydrogen, the energy output starts to fall and the core begins to contract again under gravitational forces. The energy release raises the central temperature, and at about 10^8 K the next stage in nucleosynthesis can begin. This does not happen in stars much less massive than the sun, as the gravitational energy released is insufficient to raise the temperature to the level required to overcome the Coulomb barrier between ^4He nuclei.

The obvious first stage in helium burning would be the reaction:

$$2\,^4He = {}^8Be \tag{3.20a}$$

However, this reaction is **endothermic**, and the ^8Be nucleus is unstable by rather less than 0.1 MeV. This fact—which is a reflection of the very high stability of ^4He compared with other light nuclei—is one of the reasons why heavy elements were not produced in the big bang. In stars, however, the sustained conditions of high temperature and density give rise to a small equilibrium concentration of ^8Be, which can then capture a further ^4He:

$$^8Be(\alpha,\gamma)^{12}C \tag{3.20b}$$

The combination of 3.20a and 3.20b forms the so-called **triple alpha reaction**:

$$3\alpha = {}^{12}C + \gamma \tag{3.20}$$

Because of the higher temperatures involved in helium burning, the rate of energy transport outwards from the centre of the star is greater than in the hydrogen burning phase. This high energy output causes the outer layers of the star to distend and form a large, relatively cool, shell. Thus, the star enters the *red giant* stage of its evolution. The total energy output from helium burning per unit mass of reactants is, however, only a tenth of that produced by hydrogen burning. The reaction is, therefore, completed much more rapidly and the star remains in the red giant phase for only about 1 per cent of the time that it spent on the main sequence. This is a general feature of all subsequent stages in nuclear synthesis; because of the levelling off in the curve of nuclear binding energies as the most stable nuclide ^{56}Fe is approached (see Fig. 2.2 on p. 31), the relative energy release declines. Each successive stage of nuclear burning is therefore completed more rapidly than the preceding one.

As the concentration of ^{12}C produced in the triple-alpha process builds up, further exothermic α capture reactions can take place, especially:

$$^{12}C(\alpha,\gamma)^{16}O \tag{3.21}$$

The final result of helium burning is to give roughly equal amounts of ^{12}C and ^{16}O, the most abundant nuclides in the universe after hydrogen and helium. The relative yield depends on the mass of the star. Reaction 3.21 has a higher Coulomb barrier than the previous step and so is relatively faster at higher temperatures. The heavier the star, the higher the internal temperature and so more ^{16}O is produced.

Minor reactions also occur in the helium burning phase. For example, if some ^{14}N and ^{13}C have been produced previously in the CNO cycle, we have:

$$^{14}N(\alpha,\gamma)^{18}F(v,e^{+})^{18}O(\alpha,\gamma)^{22}Ne(\alpha,n)^{25}Mg \tag{3.22}$$

and

$$^{13}C(\alpha,n)^{16}O \tag{3.23}$$

Reactions such as these may be important as a source of some nuclides such as ^{18}O. They also liberate free neutrons, which contribute to the synthesis of elements heavier than iron as described later.

When helium is exhausted in the core, the process of contraction and temperature rise sets in again, and in sufficiently heavy stars, **carbon burning** and **oxygen burning** can begin. The temperatures required to overcome the high Coulomb barriers are nearly 10^9 K, and can only be reached in stars heavier than around four solar masses. The Sun will, therefore, end its active life before this stage, throwing off some of the unprocessed outer layers and becoming a small, very dense, white dwarf. In sufficiently massive stars further elements can be synthesized by reactions such as:

$$2\,^{12}C \rightarrow {}^{20}Ne + {}^{4}He \tag{3.24a}$$

$$2\,^{12}C \rightarrow {}^{23}Na + {}^{1}H \tag{3.24b}$$

$$2\,^{12}C \rightarrow {}^{23}Mg + n \tag{3.24c}$$

$$2\,^{12}C \rightarrow {}^{24}Mg + \gamma \tag{3.24d}$$

$$2\,^{16}O \rightarrow \begin{cases} ^{28}Si + ^{4}He & (3.25a) \\ ^{31}P + ^{1}H & (3.25b) \\ ^{31}S + n & (3.25c) \\ ^{32}S + \gamma & (3.25d) \end{cases}$$

In addition to the elements formed directly, significant numbers of alpha particles, together with some protons and neutrons, are formed by these reactions. By capture reactions with other nuclei present, all the elements from Ne to Ar can be produced.

The final stage in the energy production of a star is known as **silicon burning**. This differs somewhat from the previous stages, as the temperature never rises high enough for the direct fusion of two silicon nuclei to occur. Before this point is reached, the endothermic process of **photodisintegration** becomes important. This involves reactions such as (γ,α), which is the reverse of the fusion reaction (α,γ) and occurs when the energy of the ambient γ radiation becomes sufficiently high. Free α particles are liberated, which can then be recaptured by other nuclei, in reaction sequences such as:

$$^{28}Si(\alpha,\gamma)^{32}S(\alpha,\gamma)^{36}Ar(\alpha,\gamma)\ldots \quad (3.26)$$

Finally, at temperatures above 3×10^{9} K, nuclei come into equilibrium with the α particles, and any free protons and neutrons produced in minor reactions. The **equilibrium** phase produces the elements around iron, corresponding to the maximum in the binding energy curve of Fig. 2.2. The abundance of elements in the 'iron peak' (evident in Fig. 1.7 on p. 17), therefore, results directly from the large binding energy of these elements, which leads to high equilibrium concentrations. It is only for elements near iron that the abundances can be explained as a result of equilibrium conditions. Other elements *cannot* have been formed in equilibrium, or their abundances would be much lower. Temperatures high enough for the equilibrium process to operate are in fact attained only in the central regions of a massive star. Surrounding this central core will be successive shells, containing the lighter elements produced by the other (non-equilibrium) burning reactions. Thus, the outermost part will still consist of unprocessed hydrogen, inside which will be a shell of helium produced by hydrogen burning. Further in will be a shell of carbon and oxygen, with their burning products yet closer in towards the centre. This shell structure of a heavy star at the final stage of its evolution is illustrated in Fig. 3.4.

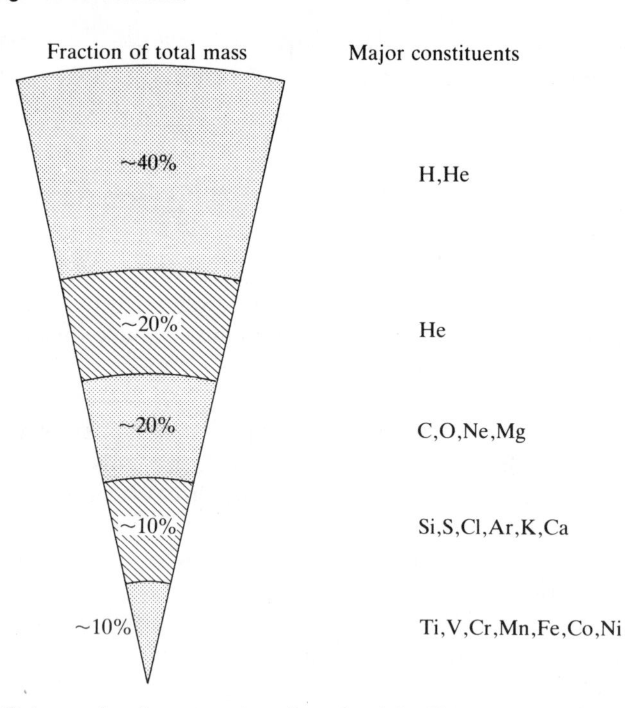

Fig. 3.4. Schematic diagram showing the 'shell' structure of a heavy star (around 25 solar masses) at the end of its evolution, just prior to a supernova explosion. The fraction of the total mass contained in each shell and the principal elements present are shown.

Supernovae and explosive burning

So far, the nuclear reactions involved in building up the elements have been considered to occur in reasonably peaceful, ordered manner. This sequence is known as **hydrostatic** burning. It takes place while the star is very nearly in a state of mechanical equilibrium, and is neither expanding nor contracting rapidly. The very possibility of such a steady state is surprising at first sight, as the rate of the exothermic nuclear reactions increases extremely rapidly with temperture. One might expect the energy output from the reactions to increase the temperature, thus further increasing the rate of energy output, raising the temperature still more, and so on, leading to a process of **thermal runaway** and the disintegration of the star in a nuclear explosion. Under normal conditions, however, there is a 'safety valve' mechanism which prevents such a runaway. Although the central density is very large, the temperature is so high that the atoms present (actually in the form of nuclei and free electrons at these temperatures) act as a nearly ideal gas. Any increase in temperature results in a rise of pressure, which causes the centre of the star to expand. The gravitational work done in lifting the outer layers lowers the internal energy, thus reducing the temperature again. Although the successive burning phases are interspersed with stages when the core contracts and heats up, the onset of a new reaction rapidly stabilizes this contraction, and it proceeds under nearly constant conditions of temperature and pressure until the reactants are exhausted.

Sometimes, the conditions for hydrostatic equilibrium in a star fail to operate. This can have various consequences. At some stages in their evolution, many stars throw off part of their outer shells. At other phases, a star may pulsate, and show a periodic change in its energy output. Many variable stars, whose magnitude varies with periods of the order of a few days, are known. The most dramatic departures from equilibrium are those giving rise to **explosive burning**, which may result in the total disruption of the star in a tremendous nuclear explosion known as a **supernova**. Such events can often be seen in distant galaxies, when the brightness of a star suddenly increases by many orders of magnitude, and then slowly fades away. It has been estimated that a supernova may occur in our own galaxy about once every 30 years on average. However, since most of the stars in our galaxy are obscured by clouds of gas and dust, nearby supernovae are not seen very often. There are historical records of such events, the best known being that recorded by Chinese astronomers in 1054. The remains of this supernova today form the Crab Nebula. The closest supernova in modern times was observed early in 1987 in the Large Magellanic Cloud, a diffuse 'satellite' galaxy of our own, visible in the southern hemisphere.

There are certainly a number of ways in which a star can explode. The mechanism which has attracted most interest is that which operates when a star runs out of energy. We have seen how in the centre of a heavy star the elements close to iron come into equilibrium at temperatures of around 3×10^9 K. When this equilibrium mixture comprising the most stable nuclei has been formed, no more energy can be obtained from further reactions. Just as when previous burning phases come to an end, the core of the star begins to contract again and the temperature rises further. This shifts the equilibrium towards lighter nuclei, and *endothermic* photodisintegration reactions (γ,α), (γ,n) and (γ,p) start to occur. The energy absorbed by these reactions accelerates the gravitational contraction and leads, rather paradoxically, to an ever more rapid rise in temperature. This, in turn, speeds up the disintegration and, in the space of a few seconds, the elements synthesized over many millions of years at the centre of the star are destroyed. Meanwhile, the surrounding regions of the star fall inwards at a rate which may approach one-fifth of the speed of light. The density at the centre increases to a value comparable to that of nuclear matter itself. Under these conditions rather surprising things can happen. The exclusion principle, which allows only two electrons to occupy each energy level, becomes very important. Electrons are forced into such high translational energy levels that it becomes favourable for them to combine with protons and form neutrons. This process, which is the reverse of the β decay of the free neutron, liberates neutrinos. At the exceptionally high energies involved these may play an important role in the transport of energy to the outer portions of the star. The collapse ends with the formation of a central core of close-packed neutrons, which cannot be compressed any further. The outer layers 'bounce back' and are ejected into space. Much of the iron from the centre of the star is predicted to be released in the form of ^{56}Ni, a nuclide formed under the very high pressures by 'squeezing' two extra electrons into ^{56}Fe. Once released into space, the ^{56}Ni undergoes β decay, first to ^{56}Co and then to the more stable ^{56}Fe. The energy output from these decay processes is partly responsible for the initial brightness of the supernova.

Many details of the final moment of a supernova are still not well understood, and this problem forms an active area of research. The 1987 observation allowed some of the theoretical predictions to be verified. Several neutrinos were detected on earth, thus providing evidence for their role in the process. However, at the time of writing the shell of expanding gas is still too hot to allow observations either of the elemental composition of the gas or of any central remnant which might be left. Observations on 'historical' supernova remnants such as the Crab Nebula do, however, give important information. Many of the elements formed by nuclear reactions can be detected by spectroscopic methods in the expanding gas clouds. Another discovery is that some supernova remnants have a **neutron star** at the centre, left behind as the collapsed core of the original star. Neutron stars can be detected as **pulsars**, as they emit radiation—light, radiowaves, X-rays, etc.—in short pulses, repeated with extraordinary regularity many times a second. This must be caused by very rapid rotation, although many properties of these objects are not well understood. It is likely that neutron stars are produced by supernovae in stars of moderate mass. More massive stars may leave behind a body which collapses totally under its own gravitational field. This is a **black hole**, where the gravitational forces are so strong that not even light can escape from the central regions. Definite experimental evidence for black holes is lacking, but many astronomers believe that there may be one at the centre of our galaxy, perhaps produced by the explosion of very massive stars in the early stages of its formation.

Supernovae can also occur in other ways, although the detailed mechanisms are still disputed. One possibility is that the 'safety valve' which controls the rate of nuclear burning may sometimes fail at the point when a new burning reaction starts. This will happen, for example, if the central density is so high that properties of the gas are controlled by the exclusion principle. Under these conditions, the pressure no longer increases with temperature as it does for a 'classical' gas. Then the temperature can rise very rapidly, leading to an explosive onset of the new reaction. This happens quite often when helium burning starts, but the resulting 'helium flash' does not destroy the star, because the temperature is rapidly brought under control again. However, it is thought by some astrophysicists that an explosive start to the next stage—carbon burning—may be more serious and can sometimes produce a supernova. Another suggestion is that in 'supermassive stars', perhaps 100 times heavier than the sun, an endothermic reaction producing electrons and positrons can become important during one of the burning stages. Just as with the photodisintegration described previously, such an endothermic process could go out of control and cause a supernova. The evidence for supermassive stars is still controversial, but they may have been common in the early stages of the formation of our galaxy.

Another important situation in which explosive burning can take place is in a system where a pair of stars orbit fairly closely about a common centre of mass. As many as 50 per cent of stars may form such binary systems and this may have an important effect on their evolution. For example, one star may reach the white dwarf stage while the other is a giant. The outer layers of the giant could then be drawn off by gravitational forces towards the other star. Falling onto the extremely dense surface of the dwarf, hydrogen and helium might be heated enough to start burning explosively. This is probably the main cause of **novae**, in which stars periodically—or sometimes at quite irregular intervals—show a large increase of brightness. However, it may be that more violent explosions, when the star system is totally destroyed in a supernova, can sometimes occur in this way.

Whatever the mechanism involved, explosive burning obviously has very important consequences. It forms the major route by which elements synthesized in the star are released into space. More subtly, it can lead to the formation of elements with relative abundances different from those produced by slow burning. Although all the elements up to iron *can* be produced by the reactions discussed in the previous section, steady-state conditions do not lead to very good predictions of relative abundances. Calculations give an over-abundance of so-called 'alpha nuclei', that is, ones such as ^{16}O, ^{24}Mg, and ^{28}Si which are 'multiples' of

^{4}He. Many of the less common nuclides with intermediate masses (especially the odd numbered ones) can be produced in their observed proportions only under conditions of explosive burning. The important factor is that very high temperatures are suddenly followed by rapid cooling as the star expands into space. The explosive process produces many less stable species, which are then 'frozen out' before they can be consumed in further nuclear reactions. This will happen not only at the centre of the star where the explosion starts, but also the outer layers which are heated in the shock wave produced. Thus, the composition of successive layers—containing the products of previous burning stages—may all be altered by the explosion.

Figure 3.5 shows the results of a recent calculation on the elements liberated in a supernova. The relative amount of each nuclide produced from a star of 25 solar masses has been compared with the observed solar system abundances. Many common elements such as iron and

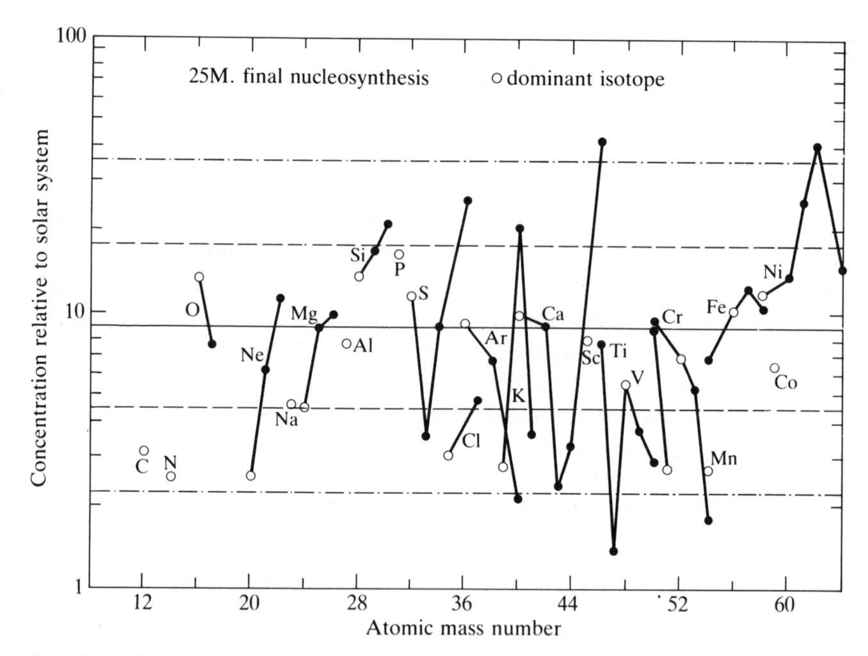

Fig. 3.5. Calculation of the elements produced by a supernova in a star of 25 solar masses. (From Woosley and Weaver 1986.) The predicted concentration of each nuclide has been compared with its observed solar system abundance. The calculation suggests that about one part in nine of the material forming the solar system has been produced by such supernovae.

oxygen are predicted to have a concentration nine times that found in the solar system. This means that their observed abundances can be accounted for if one part in nine of the material in the solar system was generated in a supernova, the remaining eight-ninths being unprocessed hydrogen and helium. There is a fair amount of scatter in the figure, but the overall agreement is impressive, especially when it is remembered that the *absolute* abundances of the nuclides plotted span a factor of more than 10^7. In fact, there is no reason to think that a calculation on a single star could explain perfectly the abundances of elements in the solar system. Stars of different masses burn at different rates, and the various reaction stages occur to different extents. For example, stars rather lighter than the one calculated will probably produce more carbon and less oxygen. The composition of the gas released will reflect these differences, and one ought to take some kind of average over the yields of stars with a range of mass. It is also important to note that supernovae are not the *only* way in which elements are released into space. Stars at some stages in their evolution may throw off part of their outer layers, either gradually over millions of years, or in a more rapid process. These outer layers will be enriched in rather different elements, for example, in nitrogen and ^{13}C produced in the CNO cycle. There is every reason to think that the gas from which the solar system formed was made up of a mixture from many different sources. No single event, therefore, can account for the abundances of the elements. In spite of these problems, the calculation illustrated in Fig. 3.5 is very encouraging and does suggest that present theories explain these abundances quite well.

The heavier elements

Following the 'iron peak', the abundances of successively heavier elements fall rapidly. These are not produced in the exothermic fusion reactions considered so far, because ^{56}Fe is the nucleus with the highest binding energy per nucleon. Elements up to two or three few places beyond iron may be produced in the equilibrium process described previously, or when silicon burns under explosive conditions. However, the majority of heavier nuclides are by-products of the major nuclear burning reactions in stars, and are generated predominantly by the process of **neutron capture**. As we saw in Chapter 2, this produces heavier isotopes of an element. Successive neutron capture events lead away from the 'line-of-stability' on the Z–N plot (see Fig. 2.10), and eventually give unstable nuclei. β decay reactions can then intervene to increase the atomic number Z, and the new element produced can undergo further neutron capture to continue the process. The abundances of heavy elements can best be explained by the operation of two distinct neutron-capture pathways, which operate in different conditions.

Free neutrons are generated as a side-product by a number of reactions in stars. Some of these occur during the helium burning stage, especially if products of the CNO cycle are present. Two reactions which are thought to be particularly important as sources of neutrons are 3.22 and 3.23. Under steady conditions of hydrostatic burning, a low flux of neutrons will be produced. Neutron capture events will be infrequent, and whenever an unstable nuclide is formed it will have ample time to undergo β decay before any more neutrons arrive. The resulting pathway is known as the **s ('slow') process**. An alternative pathway known as the **r ('rapid') process** occurs during explosive burning, when neutrons may be produced suddenly in very large concentrations. The rate of neutron capture is then much greater than that of β decay, and successive capture events can produce very neutron-rich nuclei, far off the line of stability. As the neutron flux ceases, these nuclei decay to more stable ones.

The operation of the two processes for a range of nuclides is illustrated in Fig. 3.6. The s-process pathway, shown as a continuous line on the diagram, proceeds horizontally by successive neutron capture through stable nuclides, until a β decaying nuclide is encountered. Thus, for example, we have the sequence:

$$^{182}W(n,\gamma)^{183}W(n,\gamma)^{184}W(n,\gamma)^{185}W \qquad (3.28)$$

^{185}W then decays to ^{185}Re, giving the diagonal portion of the line. The r-process, on the other hand, initially produces very unstable neutron-rich nuclei towards the bottom right-hand corner of the diagram. These undergo β decay proceses shown in Fig. 3.6 by the diagonal arrows. Some nuclei, such as ^{187}Re and ^{186}W can only be made by the r-process, as they are 'stranded' from the s-process pathway. Others, such as ^{186}Os and ^{187}Os, can only be made by the s-process, being 'shielded' from the r-process decay by the existence of the stable nuclides ^{186}W and ^{187}Re, respectively. Many nuclides, however, can be made under both r- and s-process conditions.

Since ^{209}Bi is the heaviest stable nuclide, the s-process path stops at this point, and all the heavier radioactive elements must have been produced by the r-process. Although uranium is the heaviest element surviving naturally on the earth, heavier short-lived nuclides are probably also made in supernovae. It was mentioned in Chapter 2 that some transuranium elements have been detected in small quantities in fall-out from nuclear explosions, which presumably simulate (albeit on a minor scale) the environment in which the r-process operates. There is considerable evidence that some heavy nuclei such as ^{244}Pu were present in the early solar system.

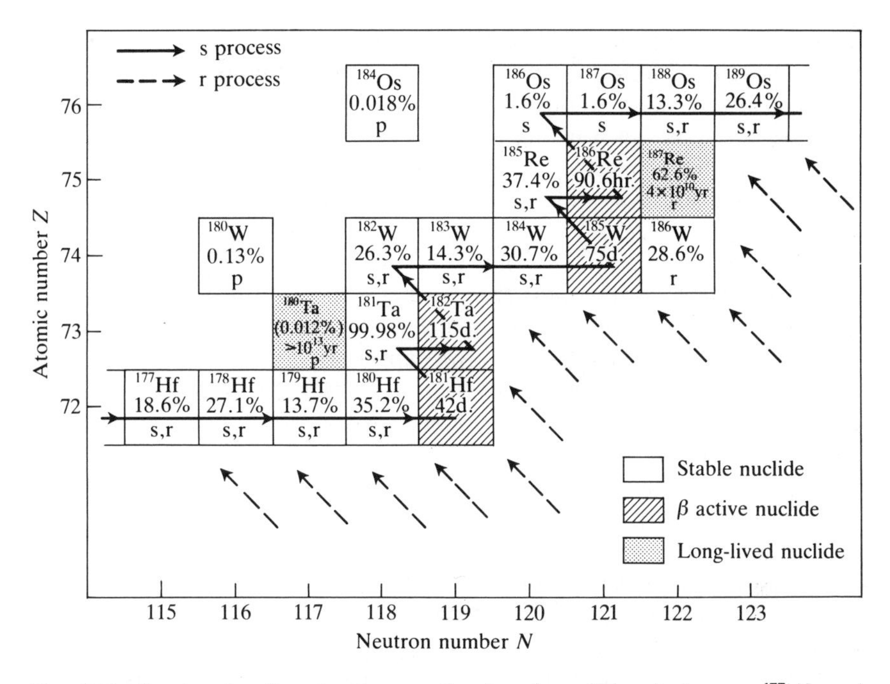

Fig. 3.6. Routes leading to the synthesis of nuclides between ^{177}Hf and ^{189}Os. The natural isotopic abundances (for stable nuclei) or half-lives for decay (unstable nuclei) are shown, together with a label p, r, or s to show the process by which each nuclide can be made. The s-process neutron capture pathway is the continuous line (——→——); the dotted lines (— — →) show the decay pathways of very neutron rich nuclei formed during the r-process.

There remain a few proton-rich nuclides, such as ^{180}W and ^{184}Os in Fig. 3.6, that cannot apparently be made by neutron capture. It is most likely that these are made by the **p-process** of proton capture, although other possibilities have been considered. The low abundance of the p-process nuclei, as illustrated with the cases of ^{180}W (0.13 per cent) and ^{184}Os (0.18 per cent) in Fig. 3.6, is very notable. It is quite likely that the p-process takes place in the outer layers of a star when they are suddenly heated and compressed by an explosive event in the core. Such outer layers still consist largely of hydrogen (protons), together with the small proportion of heavier elements that were present originally when the star formed.

Evidence for the formation of heavy elements by neutron capture can be seen in some stars. The s-process in particular may operate in the outer layers, and the spectra of a few stars show quite high abundances of the elements produced. The most prominant of these is Ba, which is made in high yield by the s-process. More remarkable is the observation of spectral lines due to Tc, a radioactive element for which the longest lived isotope ^{99}Tc has a half-life of only 2×10^5 years. Although instable itself, this nuclide forms an essential link in the s-process chain, connecting the stable isotopes of Mo and Ru through the sequence:

$$\ldots{}^{98}\text{Mo}(\text{n},\gamma)^{99}\text{Mo}(\beta^-)^{99}\text{Tc}(\beta^-)^{99}\text{Ru}\ldots$$

The strongest evidence for the two neutron capture processes comes from the observed abundances of the heavy elements. As explained in Chapter 2, the probability of neutron capture by a given nuclide depends on the **capture cross-section**. Figure 2.12 on p. 56 showed how these cross-sections vary with mass number for heavy elements. There is a strong alternation effect, nuclides with odd A having larger cross-sections than those with even A. The cross-section is also much smaller for closed-shell nuclei having a magic number of protons or neutrons. Both these patterns are reflected in the abundances of the heavy elements. If a nucleus has a high capture cross-section, it is likely to react with another neutron soon after being formed, and thus will never build up in large abundance. On the other hand, a nucleus with low cross-section will on average spend a long time before reacting further, so that its abundance will be high. This explains the alternation of abundances, elements with even atomic number (having some even A isotopes) being more common than those with odd atomic number (with only odd A stable isotopes). The magic numbers also give rise to peaks in the distribution, between Sn ($Z = 50$) and Ba ($N = 82$), and close to Pb ($Z = 82$ and $N = 126$).

A more detailed look at the pattern of abundances of individual nuclides is shown for the limited region between ^{114}Cd and ^{149}Sm in Fig. 3.7. The odd–even alternation is shown clearly, and three distinct peaks can also be discerned in the plot. One of these occurs along the line $Z = 50$ (Sn isotopes), and another along the line $N = 82$ (especially ^{138}Ba). These nuclides are made mostly by the s-process, and the peaks in abundance reflect the much lower capture cross-sections associated with the closed shells of protons or neutrons. It is particularly striking how the abundances fall sharply beyond $N = 82$, showing how much 'pile-up' occurs at this point. There is another, broader peak in the diagram, centred on the isotopes of Te and Xe, and associated with the r-process. In this mechanism, highly neutron-rich nuclides are first synthesized, which then decay along the diagonal lines with constant $Z - N$. The abundances in the r-process are therefore determined not by the capture cross-sections for stable nuclei, but those of neutron-rich ones some way from the line of stability. The Te–Xe peak can be

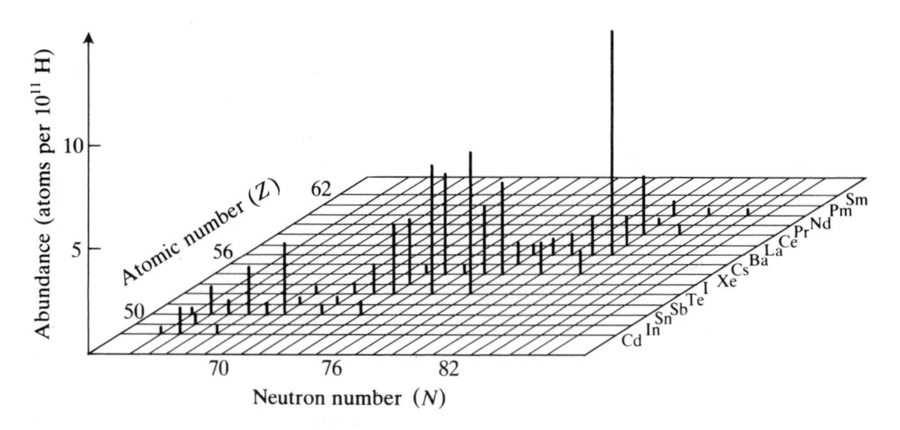

Fig. 3.7. The solar system abundances of nuclides between ^{113}In and ^{149}Sm. The plot shows the even–odd alternation of abundances, and peaks associated with the nuclear magic numbers $Z=50$ and $N=82$. As explained in the text, these features are associated with variations of neutron capture cross-section (compare Fig. 2.12 on p. 56).

explained by the decay of species with $N \simeq 50$ and $Z \simeq 82$, which will build up in high concentrations under r-process conditions because of their low capture cross-sections.

A similar pattern appears in the later abundance peak, for elements between Os and Pb. ^{208}Pb, with $Z=82$ and $N=126$, is made very efficiently by the s-process, whereas elements Os–Hg are mostly produced in the r-process, by decay of neutron-rich nuclides with $N \simeq 126$. These abundance patterns, therefore, form very strong evidence for the operation of two distinct neutron capture pathways and, taken overall, they suggest that the s- and r-processes have contributed about equally to the existence of elements beyond the iron peak. The comparative rarity of p-process nuclides, by contrast, shows that the proton capture route cannot have produced more than about 1 per cent of the total abundance.

Very light elements—the 'x' process

A few very light nuclides are not made in any of the nuclear reactions discussed so far. These are ^{6}Li, ^{9}Be, ^{10}B, and ^{11}B. (The other lithium isotope ^{7}Li is produced in the big bang, as described earlier.) In fact, these elements, far from being produced in stars, are actually predicted to be *consumed* at the temperatures involved in hydrogen burning. They are all elements of very low abundance (for example, boron is 10^7 times less abundant than carbon), but they clearly must be made somewhere. The mechanism which gives them has been called the **x-process**, and is believed to involve **spallation** reactions in cosmic rays. The origin of cosmic rays, which consist mostly of nuclei travelling at very high velocity, is uncertain. The may come from supernovae, but there are other possible processes that could accelerate particles to the energies required. Spallation reactions, mentioned in Chapter 2, occur when nuclei collide at very high energies and produce fragments of low mass number. High energy cosmic ray protons hitting heavier nuclei (or vice versa) would produce fragments including ones in the mass range from ^{6}Li to ^{11}B.

Evidence for the spallation process comes from the elemental abundances found in cosmic rays, shown in Fig. 3.8. Although these generally follow the general pattern of solar system abundances, the peaks and dips are considerably smoothed out. In particular, the elements Li, Be, and B are relatively much more common in cosmic rays.

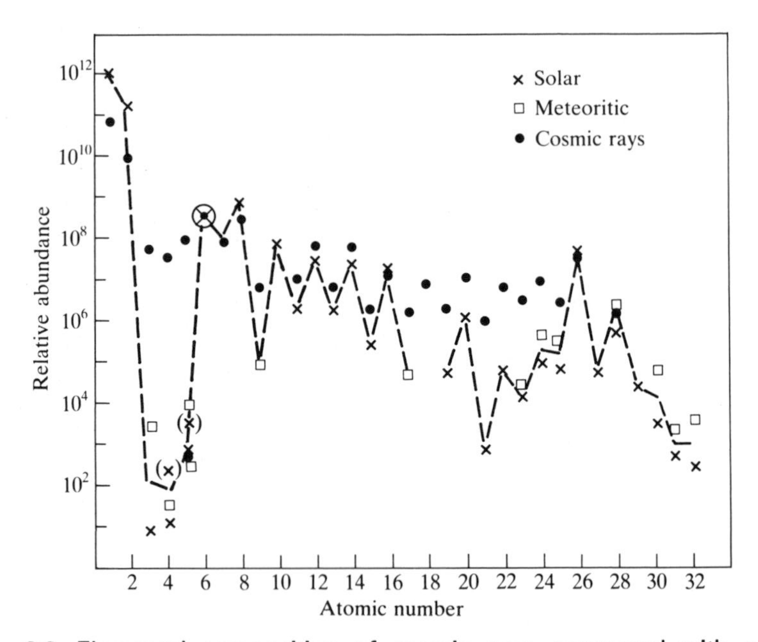

Fig. 3.8. Elemental composition of cosmic rays, compared with solar system abundances. (From Trimble 1977.) The two sets of values have been normalized at carbon. The relatively much higher abundance of Li, Be, and B in cosmic rays provides strong evidence for the origin of these rare elements.

This is consistent with the idea that the cosmic rays started with a 'normal' pattern of elemental abundance, which has been modified by spallation processes on their journey through space. Spallation must have made slight changes to the abundances of all the elements, but it is only for those nuclides that are not produced by other processes that these changes are significant.

Summary

The origin of the universe in the Big Bang about 16 billion years ago produced protons, neutrons, and electrons. Within a few minutes, about 25 per cent by weight of helium had formed, and very small amounts of deuterium and ^{7}Li. All the other elements, except for beryllium and boron, are produced by nuclear reactions in stars. B and Be, together with ^{6}Li, are the result of high energy collisions in cosmic rays.

Most of the energy of a star is produced by the fusion of hydrogen into helium. At later stages in a star's life further fusion processes take place, each stage forming heavier elements from the 'ashes' left over from the previous stage. The abundant elements carbon and oxygen are produced by helium fusion, and the elements up to iron are made in subsequent steps. Beyond iron, most elements are made by successive neutron capture, although some of the rarer nuclides are made by proton capture reactions.

Most of the elements present in the universe today come from heavy stars which end their lives with gigantic explosions—supernovae—spewing the products of the nuclear reactions out into space. As well as liberating the elements, these explosive processes also play an important role in controlling the relative abundances of the elements. The composition of the solar system suggests that it contains about 10 per cent of material derived from supernovae, the remainder being mostly unprocessed hydrogen and helium.

Further reading

The classic account of the Big Bang theory is that of Weinberg (1977). Gribbin (1987) describes some more recent developments. The astronomical background to this chapter is nicely illustrated in the *Cambridge Encyclopedia of Astronomy*, while Taylor (1970) gives a slightly more technical, but still readable treatment of the evolution of stars.

The first major review on modern ideas of stellar nucleosynthesis is that of Burbidge *et al.* (1957); although there have been many subsequent developments, this is still worth reading. Selbin (1973) gives a less detailed account, and Trimble (1977) a lengthy review of developments up to the mid-1970's. Fowler (1984) presents an interesting personal view of his own contributions in this area. Recent theoretical work on supernovae is described by Woosley and Weaver (1986). Murdin (1987) gives a preliminary report on observations of the 1978 supernova.

4
Cosmochemistry and the solar system

At the very high temperatures required for the synthesis of elements in stars, no molecules can exist; indeed, most matter is present in the form free nuclei and electrons. In the cool outer layers of some giant stars the temperature is low enough for some simple molecules to form. The best known of these is TiO, the absorption spectrum of which is characteristic of red giants with surface temperatures around 3000 K. Other molecules which have been identified from their spectra include H_2, MgH, CaH, diatomic oxides of Sc, Y, Zr, La, and Ce, and a few carbon-containing species such as CH and CN.

Although luminous bodies such as stars are the easiest ones to observe from the Earth, it is now known that a fair proportion of the matter in space is present in cooler regions, where the temperature is low enough for a wide variety of molecules and solids to form. The presence of solid dust particles has long been suspected from the observation that light from distant stars is attenuated on its way through space. More recently, a large number of molecules has been identified from measurements in the microwave region. The study of such interstellar species forms one branch of the subject of **cosmochemistry**, an interdisciplinary field that involves many types of astronomical observations as well as purely chemical aspects. One of the major challenges in this area, however, is to understand the chemical composition of the Earth and of the other bodies that make up the solar system. The discussion of this problem forms the larger part of the present chapter. The solar system is believed to have been formed at the same time as the Sun, around 4.6 billion years ago. Although important traces still remain of the chemical reactions that took place then, most of these have long since ceased. On the other hand, the chemical activity found today in some regions of interstellar space may be similar to that involved in the early stages of the solar system. It is partly for this reason, as well as for its intrinsic interest, that this topic is discussed first.

Interstellar gas and dust

The average density of matter in interstellar space is around 10^6 atoms per m^3. The major species present are hydrogen and helium atoms and, indeed, the microwave spectrum of the hydrogen atom, which gives a line at 21 cm wavelength, forms a useful way of mapping the distribution of interstellar matter in our galaxy. Any molecules forming under these conditions will be rapidly dissociated by high-energy ultraviolet (UV) photons from stars, and for this reason it was once thought that complex molecules would be unlikely to exist in space. In recent years, however, many molecules—of which a selection is given in Table 4.1—have been detected. Most of these are found in clouds of relatively 'high' density, ranging from 10^8 atoms per m^3 in **diffuse clouds**, up to 10^{10} per m^3 in **cool molecular clouds**. In diffuse clouds only simple species are present,

Table 4.1
*Molecules observed in interstellar space**

Diatomics			
H_2	CH^+	CH	OH
C_2	CN	CO	CO^+
NO	CS	SiO	SO
NS	SiS		
Triatomics			
H_2O	C_2H	HCN	HNC
HCO^+	N_2H^+	H_2S	HCS^+
OCS	SO_2	NaOH	
Four atoms			
NH_3	C_2H_2	H_2CO	HNCO
C_3N	H_2CS	HNCS	
Five atoms			
CH_4	CH_2NH	CH_2CO	NH_2CN
HCOOH	C_4H	HC_3N	
Six atoms			
CH_3OH	CH_3CN	NH_2CHO	CH_3SH
Seven atoms			
CH_3NH_2	CH_3C_2H	CH_3CHO	CH_2CHCN
HC_5N			
Eight atoms			
$HCOOCH_3$			
Nine atoms			
C_2H_5OH	$(CH_3)_2O$	C_2H_5CN	HC_7N
Eleven atoms			
HC_9N			
Thirteen atoms			
$HC_{11}N$			

*From Duley and Williams (1984).

and can be identified by their absorption lines in the visible or UV spectrum. The denser molecular clouds on the other hand are quite opaque to short wavelength light and this is what protects the molecules in their interior from dissociation by UV radiation. Molecules in these clouds have mostly been identified from emission lines in the microwave region, accessible to radio telescopes.

Many of the molecular species listed in Table 4.1 are well known in the laboratory. For example H_2, although difficult to observe, is by far the dominant molecule present, and another common species is CO. Some of the molecules found in space, however, are not known on Earth. Especially striking is the series of cyanoacetylene molecules $H{-}(C{=}C)_n{-}CN$, with n up to 5. The spectroscopic lines by which these molecules are identified arise mostly from transitions between rotational levels. In cases where molecules are unknown in the laboratory, such identification must be aided by theoretical calculations.

The presence of a wide variety of molecules in interstellar clouds raises several difficult problems. In the first place, many of the more complex species are stable only under relatively cool conditions. The temperature in interstellar clouds can be found by measuring the intensities of different absorption and emission lines, which give information about the relative numbers of molecules in ground and excited states. For example, if two states, with an energy difference ΔE, have n_1 and n_2 molecules in them respectively, then according to the Boltzmann distribution

$$n_2/n_1 = \exp(-\Delta E/kT) \qquad (4.1)$$

where k is the Boltzmann constant (1.38×10^{-23} J/K). Temperatures estimated in this way range from 40 K in cool molecular clouds to a few hundred degrees in diffuse clouds. Matter is originally expelled from the surfaces of stars at a temperature of some thousands of degrees and even out in space it is still subject to the heating effects from radiation. Therefore, it is important to understand first of all how interstellar gas loses energy so that it can cool to low temperatures. Undoubtedly, the most important mechanism is by radiation from excited atoms and molecules. The reaction scheme below shows how a collision between two species, A and B, can convert translational kinetic energy into internal energy, so giving an excited state B*, which then emits energy by radiation of a photon ($h\nu$):

$$A + B = A + B^* \qquad (4.2)$$

$$B^* = B + h\nu. \qquad (4.3)$$

The type of excited state involved in such a process must depend on the temperature range. At around 10^4 K excited electronic states of many atoms can be produced and, because of their high abundance, excited H atoms form the major source of radiative energy loss. In the 10^3 K range excited states of H are unaccessible, and it is probably electronic states of species such as Si^+ and Fe^+ which are most important. At lower temperatures, the energies involved are too low to populate excited electronic states, and energy is lost predominantly by emission of infra-red and microwave energy from the rotational and vibrational states of molecules such as CO. The rate of the spontaneous emission process 4.3 is, however, proportional to the cube of the photon energy $h\nu$ and so, at lower temperatures, the rate of cooling through radiation is very slow.

A second important problem concerns the way in which molecules form. It has already been pointed out that most complex molecules are found in dense clouds where they are largely protected from dissociation by UV radiation. Even under these conditions some dissociation must occur, and the concentration of each species depends on a balance between the rates of formation and destruction. The fastest reactions are those where an ion is involved, for example:

$$C^+ + H_2 = CH^+ + H. \tag{4.4}$$

This type of process is rapid because the charged ion such as C^+ polarizes the H_2 molecule, inducing a dipole moment which leads to a long-range attraction between the two species. Molecular ions like CH^+ can undergo further reactions, which probably form the major route to the production of organic molecules in interstellar space. The most difficult step, however, is the initial production of H_2 molecules from hydrogen atoms. A diatomic molecule cannot form from the collision between two atoms, unless there is some mechanism by which the kinetic energy of the colliding species may be lost. Figure 4.1 shows this, and indicates some of the possible mechanisms for molecular formation. Collision between the two atoms leads to an unbound state which will immediately dissociate again (Fig. 4.1(a)). Energy could be removed by the presence of a third atom during the collision (Fig. 4.1(c)), but under the very low pressures existing even in relatively 'dense' clouds in space, such three-body collisions are extremely rare. Another possibility is that of radiative decay, in which the colliding system reaches a bound state by emitting a photon (Fig. 4.1(b)). Since the duration of the collision is around 10^{-13} second, however, this is again an unlikely process. Although it may be an important mechanism for the production of some diatomic molecules, it cannot explain the rate of formation of H_2. The most promising source of this is through an entirely different mechanism, that of surface processes occurring on dust particles (see Fig. 4.1(d)). H

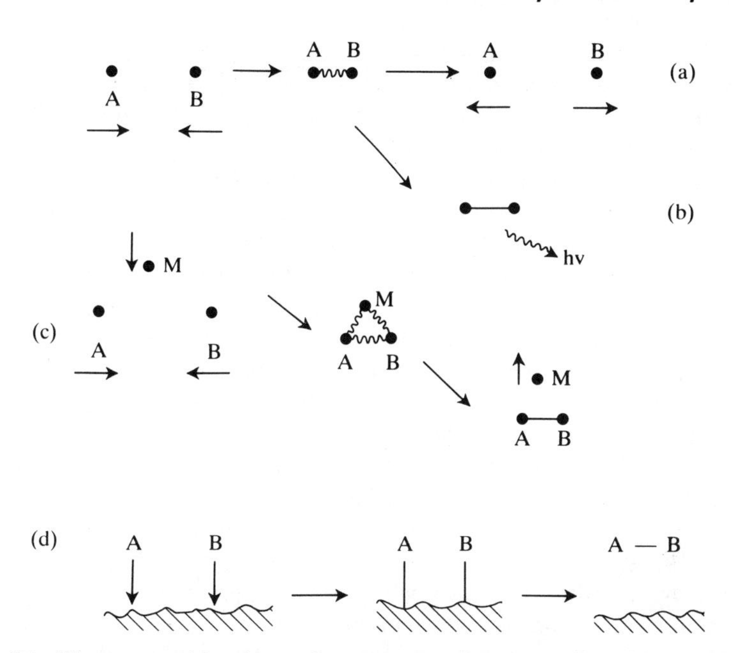

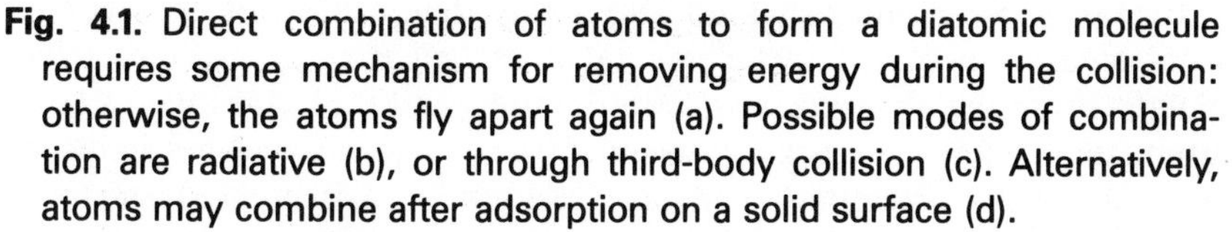

Fig. 4.1. Direct combination of atoms to form a diatomic molecule requires some mechanism for removing energy during the collision: otherwise, the atoms fly apart again (a). Possible modes of combination are radiative (b), or through third-body collision (c). Alternatively, atoms may combine after adsorption on a solid surface (d).

atoms can adsorb on these particles, and combine on the surface to form molecular hydrogen, which then desorbs.

It is worth noting that the problems of formation of polyatomic molecules by collisions between smaller species are less severe than for diatomic molecules. In a reaction such as

$$A + CB = ABC \tag{4.5}$$

the energy released in forming the new bond can be partly redistributed into the other bond or bonds in the new molecule. Although this energy must be lost fairly rapidly to avoid dissociation, the time-scale for this process can be much longer than in a simple collision between two atoms. Thus, three-body or radiative mechanisms for energy loss can operate much more easily.

As mentioned above, the presence of dust particles in space can be inferred from the attenuation of light from stars. The average density of such particles is very low, around 10^{-6} per m^3, but they are particularly interesting as they contain the kind of material from which the planets formed, as described in the following section. Unfortunately, it is not easy to determine the nature of this solid material. Certain broad absorption features of interstellar dust can be seen in the infra-red region of the spectrum. Not only do solids have much broader absorption bands than those of molecules, but the spectra also depend on the size of the solid particles, which are probably of the order of 1 μm and, hence, comparable to the wavelength of the radiation. In spite of these difficulties, it is believed that some clouds contain particles of carbon (in the form of graphite) and SiC. Silicates such as $MgSiO_3$ and metallic iron are probably also present. Whereas these latter solids are important constituents of the solar system, solid carbon and carbides do not play an important part in most of the models of the chemistry of the early solar system. As will be discussed later, it appears that the relative abundance of carbon and oxygen is crucial in determining which solids form. Many of the dark clouds containing graphite and complex organic molecules are in carbon-rich regions where the [C]/[O] abundance ratio is higher than in the solar system. These clouds probably form from material released by certain giant stars which are known to have an especially carbon-rich surface at a late stage in their evolution.

Evidence of an indirect kind about the composition of dust comes from measurements of the elemental abundances in some interstellar clouds. From the absorption spectra of atoms and molecules the relative abundances of the elements present in the gas phase can be deduced. Although many variations of detail are found, the lighter non-metallic elements such as H, C, N, and O show relative abundances close to their solar system values. Many of the metallic elements and also silicon seem, however, to be depleted. These depleted elements are precisely the ones which would be expected to form the most stable solid phases. Figure 4.2, for example, shows the abundances of elements found in a typical cloud. The gas-phase abundances, relative to the solar system values, are plotted against the predicted condensation temperatures for the different elements, as discussed later (see Fig. 4.4). The good correlation shows that the most depleted elements are those that are expected to condense to the solid phase most easily. Although these observations do not show what actual compounds are present in the solid grains, they offer strong support to the theoretical models of chemical condensation of the elements, which play an important part in current ideas about the formation of the planets.

Another problem concerning interstellar dust is the mechanism of its formation. Although the condensation of many solids is strongly favoured thermodynamically at temperatures below 1500 K, there is an important kinetic barrier to the nucleation of solid grains, especially at low pressures. This problem, indeed, is much more severe than that of

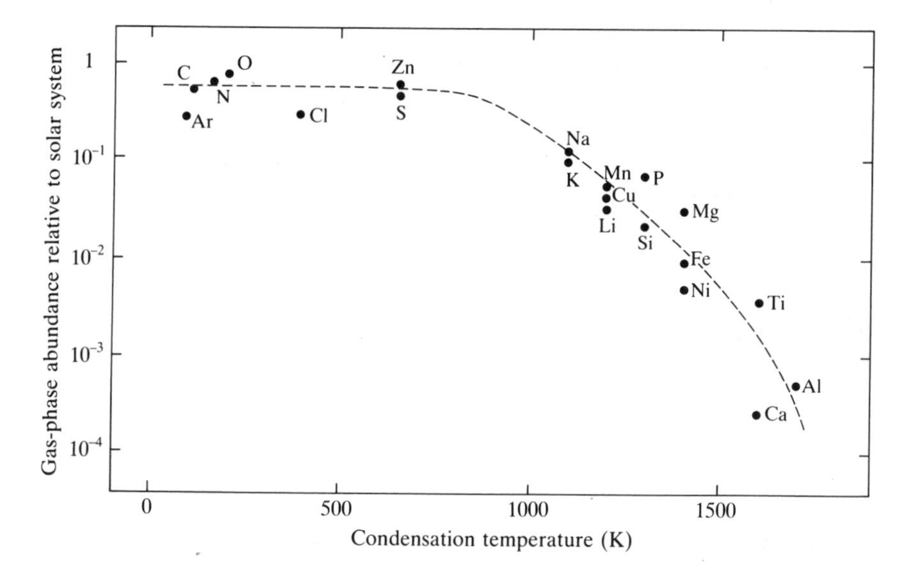

Fig. 4.2. The abundances of elements in the gas phase found in the ζ Ophiuchi cloud, relative to solar system values, plotted against predicted condensation temperatures (see Fig. 4.4). The trend shown by the dashed line suggests that involatile elements are removed from the gas phase by condensation into solid grains. (Data from Duley and Williams 1984.)

the formation of molecules. Calculations show that solids cannot nucleate at an appreciable rate even in relatively dense molecular clouds. It is much more probable that the initial formation of solids occurs at higher pressures soon after the elements have been expelled from stars. Cooling by radiation is quite efficient in the temperature range of 10^3–10^4 K. Thus, the gas can cool to below the condensation temperature of 1500 K while the density is still high, perhaps 10^{14} atoms per m^3. We shall see later that some of the solids which make up the planets may have formed in this way, quite close to the stars from which the elements themselves originated.

The origin of the planets

A number of hypotheses have been put forward to explain the origin of the planets. According to some theories another star passed near the Sun, and the planets were formed from material either drawn out of the Sun by gravitational forces or captured by the Sun from the visitor.

Strong arguments can now be found against these ideas, and most current models are based on the **nebula hypothesis** originally suggested by Kant and Laplace. According to this theory, the planets formed from a gas cloud or 'nebula' which surrounded the Sun as it formed. The hypothesis is supported by the recent observations that very young stars do, indeed, have such clouds around them. Thus, planetary systems may arise quite naturally during star formation, although this does not necessarily mean that all stars have planets. Many stars form systems where two or even more stars orbit each other fairly closely, and under these conditions, it is doubtful whether a stable planetary system could exist.

The formation of the planets must have required both gravitational and chemical forces. Large bodies such as stars can be formed by the action of gravity on a gas cloud, causing it to contract and heat up as described in Chapter 3. Gravitation alone, however, is not sufficiently strong to hold together smaller masses of gas from which the planets could have formed and so chemical condensation into solid particles must have happened first. On the other hand, chemical forces alone cannot explain the formation of bodies larger than a few centimetres across. Thus, the planets must have come from the **accretion** of dust through gravitational forces.

To understand in detail how gravitational forces led to the accretion of the planets from the solar nebula is a formidably difficult problem in mechanics. Many aspects are still uncertain, but enough progress has been made in recent years to follow the broad outlines of the process. The solar system as a whole formed by the gravitational contraction of a large mass of gas. Probably, this was part of a cloud large enough to form a cluster of several stars, which have since become widely scattered. The contraction may have been initiated by shock waves emanating from supernova explosions of older stars in the vicinity, although this is not an essential part of the theory. Gravitational attraction led to a build-up of density in the central part of our solar system, from which the Sun developed. This central concentration was surrounded by a more diffuse cloud of gas and dust, the nebula from which the planets accreted. The entire planetary system could have been made from material with a mass only 1 per cent of that of the Sun, but most theories suggest a nebula of about 0.05 solar masses. A slow rotation of the nebula caused it to form a flattened ellipsoid shape as it contracted, as shown in Fig. 4.3. Frictional forces on the dust particles, due to their motions through the gas in the cloud, will have caused these to settle quickly into the plane of rotation, as in Fig. 4.3(b). The same effect keeps the rings of Saturn and some other planets in a flat disc configuration. These ring systems probably formed by the break-up of satellites by strong tidal forces in the

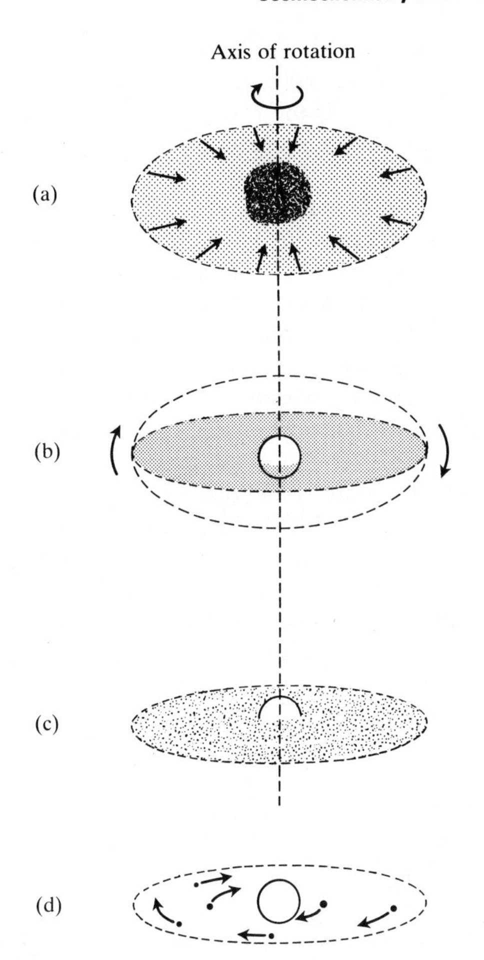

Fig. 4.3. Probable sequence of steps in the formation of the solar system. (a) Gravitational contraction of a rotating gas cloud leads to a dense central region (eventually forming the Sun) and a more diffuse, flattened nebula. (b) Dust particles from the nebula settle into a disc. (c) Accretion of dust into numerous small **planetisimals**, each a few kilometres in diameter. Collisions between planetisimals lead to capture, disintegration, or deflection of their orbits. (d) Eventually larger bodies capture the smaller ones. Uncondensed gas is blown away by the 'solar wind'; this process may begin in earlier stages.

gravitational pull of their planets, but in the case of the much more extended solar nebula the reverse would have happened. Small concentrations of dust, initially occurring by chance, attracted more material towards them, and most of the solid material would have aggregated into bodies a few kilometres in diameter, known as **planetisimals** (Fig. 4.5(c)). The next stage is the formation of larger planets from these small bodies, but this was a more complicated process, and certainly took a much longer time. Two planetisimals could fuse together in a collision, but there were many other competing effects. A collision occurring with a high relative velocity would be more likely to disrupt the planetisimals into smaller fragments. Near-collisions would also deflect planetisimals from their originally circular orbits into more eccentric ones. In this way some planetisimals must have escaped altogether from the solar system; others may have come too close to the Sun and thus evaporated. Eventually, however, a few larger bodies formed which attracted and absorbed the smaller ones.

Calculations suggest that the final stage of accretion was mostly completed in about 10^8 years. Some planetisimals continued to roam the solar system for long after this, before most of them were eventually captured. Some of the satellites of Jupiter and Saturn, photographed in the Voyager spacecraft missions, have surfaces completely covered by impact craters, presumably caused by infalling plantetisimals late in the accretion stage. Many craters also survive on the moon, although much of its surface is younger than those of the outer satellites. Most of the planets themselves have surfaces that have been melted and changed by other geological processes, and direct evidence of the accretion stage has been lost. Even today the accretion is not quite complete. Meteorites, comets, and asteroids are all examples of small bodies in the solar system which have failed to form larger planets. For this reason, they can give important information about the early solar system, which will be described later.

Another important ingredient of the theory is the loss of uncondensed gas from the solar system. This would start to happen when the Sun became active. The solar system today is pervaded by energetic charged particles (nuclei and electrons) which 'evaporate' from the outer layers of the Sun. It appears from observations of young stars that such a **solar wind** may be much stronger in the early stages of star formation. The strong solar wind would be sufficient to blow away the remaining gas, although it is difficult to estimate how quickly this happened. It is most likely that the loss of gas occurred after the formation of the planetisimals and during the early stages of their accretion into planets.

The condensation of the elements

An essential feature of the model just described is that the planets formed from solid dust particles. To understand the chemical constitution of the planets, therefore, it is necessary to know the composition of this dust under the conditions of temperature and pressure appropriate to the solar nebula. In principle, this could be investigated experimentally by making a mixture containing the correct proportions of all the chemical elements. Starting at a high temperature in the gas phase, one would then slowly cool the mixture and look at the composition of the solids present at each temperature. (Intermediate condensation to the liquid phase is unlikely, because of the low pressure of the nebula.) Such an experiment would be very difficult to perform realistically in the laboratory, and most of the detailed information on the condensation process has come instead from 'computer experiments', that is, from theoretical simulations of the cooling sequence just described. Using thermodynamic data for a large number of possible gas-phase and solid species, the equilibrium constants for the formation of the different solids can be calculated as a function of temperature, and it is possible to determine how the elemental composition of the solids should change as the temperature is lowered. Of course, the calculations also require an estimate of initial composition of the mixture and its total pressure.

The results of a typical calculation are shown in Fig. 4.4. The elements are assumed to be present in their solar system abundances as measured today and the total pressure is taken as 10^{-4} bar (10 Pa). Condensation starts at temperatures above 1600 K, and continues down to the lowest temperatures accessible in the solar system. All elements condense over a certain range of temperature and (except for oxygen, where this range is especially wide) Fig. 4.4 shows an average temperature at which about half of each element is in the solid phase.

It is useful to divide the condensation sequence into a number of stages, indicated in Fig. 4.4.*

1. The fraction condensing above 1400 K is known as the **early condensate**. It comprises mixed oxides of Ca, Ti, and Al and some other elements, together a metallic phase containing heavy transition elements.

2. The **metal + silicate** fraction is made up largely of Mg silicates together with metallic Fe and Ni. Because of the relatively high abundance of the elements in this fraction, it contains well over half of the total mass of solid condensing above 200 K.

3. The group of **volatiles > 600 K** includes alkali metals which enter the silicate phase, and the relatively abundant element S which combines with metallic Fe at about 650 K by the reaction:

$$Fe + H_2S = FeS + H_2. \quad (4.6)$$

*Although this kind of grouping is widely used, the results of different calculations give slightly different values for the temperature ranges and the elements involved. For this reason, the classification scheme adopted here (based on condensation temperatures quoted in Ringwood (1979)) differs slightly from that found in some of the other references given at the end of the chapter.

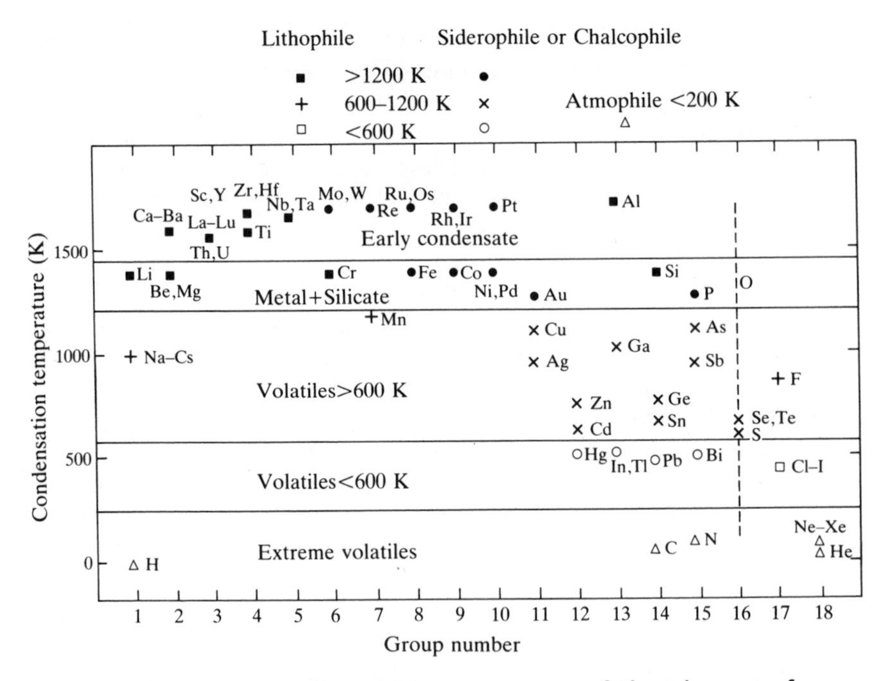

Fig. 4.4. Predicted condensation temperatures of the elements from a nebula with a total pressure of 10^{-4} bar (10 Pa). The symbols show the predominant solid phase formed and the temperature range of condensation. (From data in Ringwood 1978.)

A number of other elements also enter the metal-sulphide solid solution in the temperature range 600–1200 K.

4. **Volatiles < 600 K** include some metals which condense as sulphides, and halogens which form compounds with very electropositive metals.

5. A final category of **extreme volatiles** includes the noble gases and elements left as volatile hydrides. At temperatures down to 20 K, a majority of these species condense to form **ices**, but hydrogen and helium are left in the gas phase. Because of the very high abundance of H and He in the solar system, this residual gas in fact comprises 98 per cent of the total mass; the ices represent about 1.5 per cent and the 'rock' fraction condensing above 300 K only 0.5 per cent of the mass of the original gas.

The wide condensation range of oxygen is a reflection of its high abundance, and of the large variety of compounds that it forms.

Condensation starts with the oxide phases of the early condensate and the silicate fractions, but much oxygen is left at lower temperatures in the form of water. This forms ice at 150 K, but some water can be taken into the solid phase before this, by the formation of hydrated silicate minerals in the temperature range 300–400 K. Another important reaction is the oxidation of iron. Some Fe(II) should enter the silicate phase at around 600 K, and the reaction:

$$3/4\,Fe + H_2O = 1/4\,Fe_3O_4 + H_2 \tag{4.7}$$

occurs at about 400 K. Thus, the average oxidation state of iron in a planet should depend on the temperature at which condensation occurred. This is important in some theories of the formation of the planets.

In addition to showing the condensation temperature, Fig. 4.4 indicates the type of solid phase in which the element is found, using the geochemical classification of Fig. 1.5 (see p. 13). There is a strong correlation between the chemical form in which elements are found on the Earth today and the way in which they are predicted to condense. Thus, lithophile elements mostly condense as oxides, siderophiles in the metallic phase, and chalcophiles as sulphides, although it must be emphasized that the stability of these different compounds may vary with temperature, as with the case of Fe discussed above.

The detailed calculations leading to Fig. 4.4 are very complicated, but the general trends can be understood fairly simply by looking at some basic thermochemical data, which is shown in Fig. 4.5. Figure 4.5(a) shows the heat of vaporization of the solid elements, either to form atoms, or (in a few cases) molecular species. Figure 4.5(b) gives the heat of formation of the oxides of the elements, per mole of O used. In each case, the scale is arranged so that elements appearing higher in the plot should have more tendency to enter the solid phase. It can be seen that the early condensate fraction consists largely of those elements with the largest heat of atomization (condensing as metals) and those with the most stable oxides (which, indeed, condense in this form). The metal + silicate fraction contains some oxides slightly lower on the corresponding plot Fig. 4.5(b) and metals with heats of sublimation around 400 kJ/mol in Fig. 4.5(a). With a few exceptions, elements condensing below 1200 K are ones which are both lower than Si on the oxide plot, and lower than Ni in Fig. 4.5(a). Thus, the relative stabilities of both the oxides and the solid elements play a dominant role in controlling the cosmochemistry and the geochemistry of an element. Generally speaking, the lithophile elements are those where the heat of formation of the oxide is more favourable than that of Fe. (The halogens, occurring as metal halides, are also classified as lithophiles, as they are

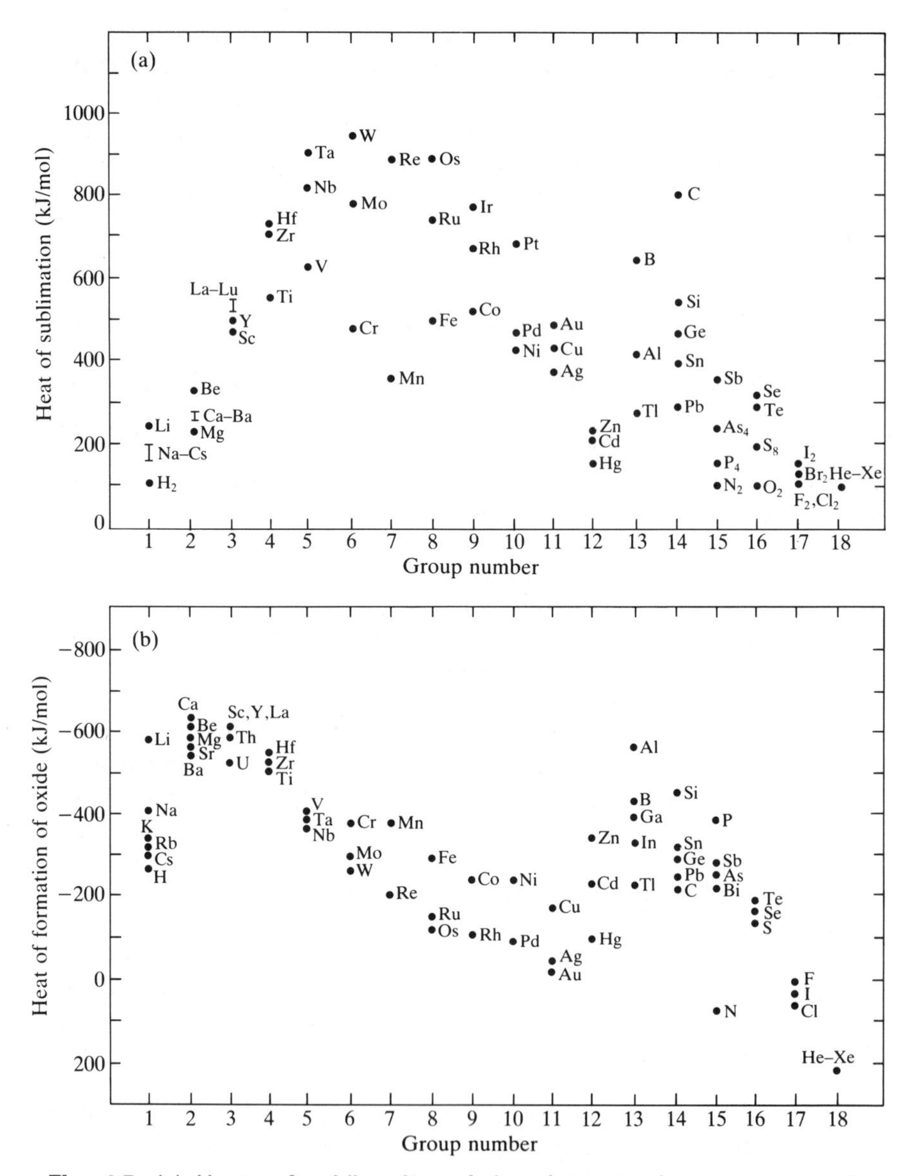

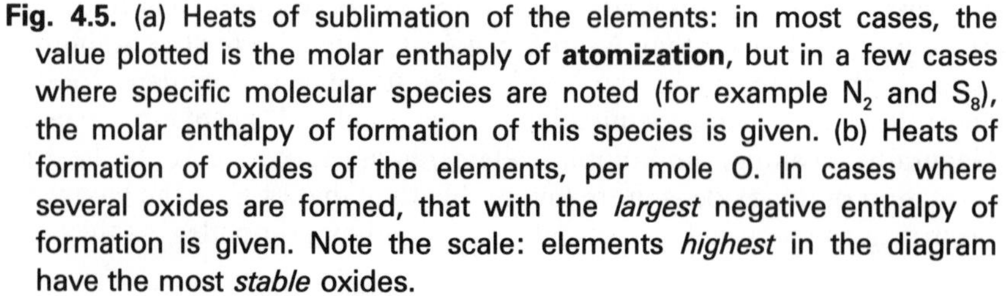

Fig. 4.5. (a) Heats of sublimation of the elements: in most cases, the value plotted is the molar enthaply of **atomization**, but in a few cases where specific molecular species are noted (for example N_2 and S_8), the molar enthalpy of formation of this species is given. (b) Heats of formation of oxides of the elements, per mole O. In cases where several oxides are formed, that with the *largest* negative enthalpy of formation is given. Note the scale: elements *highest* in the diagram have the most *stable* oxides.

also found in ionic compounds in the outer layers of the Earth.) The siderophiles which enter the metallic phase of the core are involatile and rather inert elements. The relatively high abundance of the non-metallic element sulphur is also important, as it controls the condensation of the chalcophile elements. These are mostly metals in the groups 12–15 in the periodic table, and are noted for their relative affinity for sulphur and other non-metals of low electronegativity. Finally, the atmophiles are elements which are either chemically inert, or which have stable volatile hydrides.

Carbon, the fourth most abundant element in the solar system, is unusual in its behaviour. It has a large heat of sublimation, and as a pure element would be very involatile. The predicted condensation temperature is low, however, largely a consequence of the volatility of its oxides, especially CO under the conditions of the solar nebula. Below 500 K the equilibrium:

$$CO + 3H_2 = CH_4 + H_2O \qquad (4.8)$$

moves to the right, so that most carbon is predicted to condense finally as methane. In the gas phase, however, reaction 4.8 is extremely slow. Laboratory studies of the **Fischer–Tropsch** reaction of CO and H_2 show that it is catalysed by a variety of metal and other surfaces, which would certainly have been present in the dust of the solar nebula. In addition to the equilibrium product methane, a variety of heavier hydrocarbons can be produced. Such heavy hydrocarbons are less volatile than methane, and as we shall see later, they are found in some meteorites.

The possible formation of thermodynamically less stable products such as complex hydrocarbons illustrates one limitation of theories based on the **equilibrium** condensation of elements from the gas phase. It is also important to know the effect of changing the initial conditions of composition and pressure in the solar nebula. Increasing the pressure raises the condensation temperatures of all the elements, but it is found that the Fe-rich metal phase is relatively more affected by this than the silicate phase. Whereas at 10^{-4} bar total pressure the main metal and silicate fractions condense at the same temperature, at 10^{-1} bar the Fe will solidify first. Although it has been suggested that the inner Fe core of the Earth may have condensed before the surrounding silicate mantle, the pressure required for this is probably unrealistically high, and it is generally believed that the metal and silicate fractions did condense at very similar temperatures.

The composition of the nebula is also critical, and the calculations show that the most important parameter here is the ratio between carbon and oxygen concentrations. The results summarized in Fig. 4.4 come from a [C]/[O] ratio of 0.6, which is the current solar system value measured from the Sun's spectrum. Although it is difficult to see how the relative abundances of C and O could have changed since the origin of the solar system, it is quite possible that the composition of the solar nebula was not fully homogeneous, and that there may have been regions with different [C]/[O] values. With [C]/[O] = 0.6, CO is the predominant form of carbon at the temperatures where most elements condense. Thus, there is excess oxygen present, to form oxides of the lithophile metals. If [C]/[O] were larger than 1.0, however, nearly all the oxygen would be combined as CO. The partial pressure of free oxygen would then be many orders of magnitude lower, with the result that the condensation temperatures of oxides would be considerably reduced. Calculations show that the first solids to condense under these conditions are not oxides, but compounds such as SiC, CaS, TiN, and Fe_3C, which form at temperatures around 1500 K. The major metal phase (dominated by Fe) is unaltered in its condensation range of 1300–1400 K. Oxides do not start to condense until around 1200 K and magnesium silicates at 1100 K. It is somewhat alarming that a relatively small change in the [C]/[O] ratio can have such a large effect on the early stages of the condensation process. The more 'oxidizing' [C]/[O] ratio of 0.6 probably represents better the conditions under which most of the solar system condensed, but it is possible that there may have been regions of the nebula with a more carbon-rich composition.

The composition of the Earth

A complete theory of the formation of the planets should be able to predict *ab initio* their chemical composition. Unfortunately, none of the theories so far developed can be trusted to give sufficiently detailed information about the conditions of temperature and pressure prevailing in the solar nebula to do this. The whole process, indeed, was so complicated that it is doubtful whether such a goal will ever be completely realized. A more fruitful approach is to look at the actual composition of the Earth and other bodies in the solar system, and to see what information this can give about how they were formed. It is appropriate first to look at the Earth; other parts of the solar system will be discussed later.

The abundance of elements in the Earth is clearly determined by two quite different factors: the composition of the nebula from which the solar system formed, and the proportion of each element which condensed into the solid phase. The composition of the nebula—which depends on the processes of nuclear synthesis described in Chapter 3—is generally assumed to be the same as that found for the present-day solar system. To see what proportion of each element condensed, therefore, one should look at its abundance on Earth *relative* to the overall solar system value. Such abundance ratios for the elements, using the estimated composition of the Earth from Fig. 1.7, are shown in Fig. 4.6. The scale has been normalized to make the abundance ratio of silicon equal to one; this is more convenient than using 'absolute' ratios, as hydrogen—the most abundant element in the solar system—is comparatively very rare on Earth. Any element which condensed to the same extent as Si should plot at unity on the vertical axis, and elements below the horizontal line in Fig. 4.6 must have condensed less efficiently from the solar nebula.

Each element in Fig. 4.6 is plotted with a symbol showing both its volatility range and the phase in which it condenses. The symbols are the same as those used in Fig. 4.4, and it is instructive to compare these two plots. Although there is a fair amount of scatter in the relative abundances, the values correlate strongly with the condensation

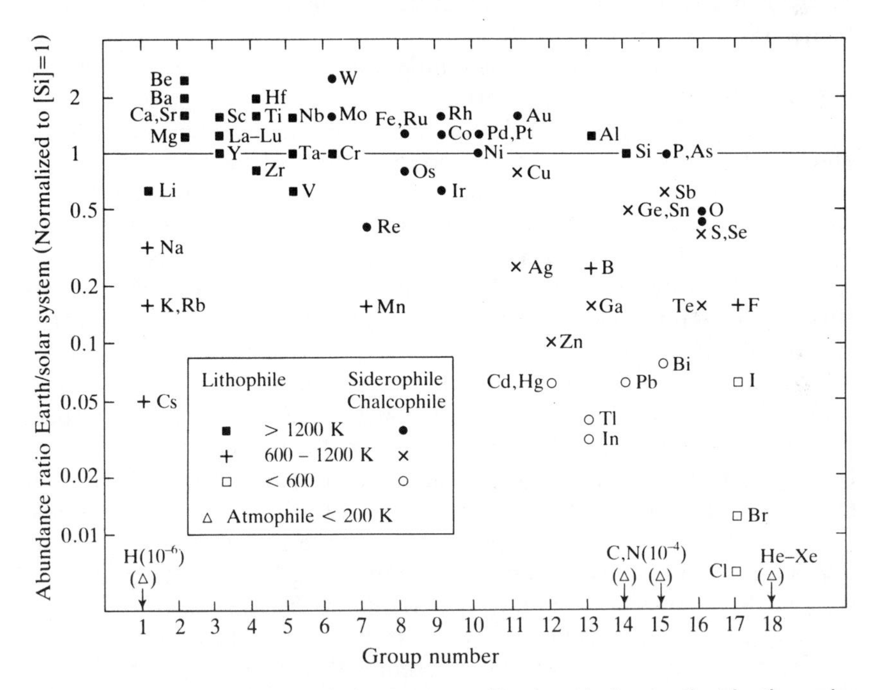

Fig. 4.6. The abundance of elements on Earth, relative to that in the solar system. The logarithmic scale is normalized to Si=1. The symbols, which show the volatility range and geochemical behaviour of each element, have the same meaning as in Fig. 4.4, with which this plot should be compared.

temperatures. Elements in the same condensation range as silicon—that is, in the 'metal + silicate' fraction—have a very similar relative abundance, clustering around unity on the plot. The 'early condensate' fraction of elements, predicted to condense above 1400 K, is slightly enriched relative to silicon, whereas the more volatile elements are depleted, those in the 600–1200 K range by up to a factor of 10, and ones condensing below 600 K by rather more than this. All the extremely volatile elements, forming the ice fraction at very low temperatures, have terrestrial abundances more than three orders of magnitude less than in the solar system as a whole.

The scatter found in Fig. 4.6 may come from a variety of sources. We emphasized in Chapter 1 that models of the Earth's total composition are subject to many uncertainties, reflecting an incomplete knowledge of the core and mantle. It has often been suggested that cosmochemical theories of element condensation might provide a better guide than purely 'empirical' estimates and, in fact, the abundances of seven elements were 'adjusted' in this way to obtain the composition used in this book (see Appendix A). However, it must be remembered that theoretical modelling also has its limitations. Some of these have already been mentioned, and will be enlarged on later.

Accepting the correlation displayed in Fig. 4.6, between the relative abundance of an element and its condensation temperature, what can one now say about the conditions under which the Earth formed? The simplest answer would be that the Earth accreted from solids which condensed from the solar nebula at a temperature of about 600 K. Qualitatively, one can then understand the strong depletion of elements still volatile at this temperature, and the higher abundance of less volatile elements. Unfortunately, this simple picture cannot be quite correct. At 600 K virtually no hydrogen should be present in the solid phase, as water (the main form of hydrogen on Earth) does not condense to form hydrated silicates above 400 K. If one lowers the condensation temperature sufficiently to account for the presence of water, it is then impossible to understand the strong depletion of less volatile elements, even those in the '< 600 K' fraction. There are also strong indications that the different zones of Earth are not in internal chemical equilibrium with each other. For example, a siderophile element such as nickel concentrates in the metallic core, but it is also found in oxide minerals in the crust, with an abundance much greater than would be expected on thermodynamic grounds. Further evidence of disequilibrium comes from the oxidation state of iron. This occurs in metallic form in the core, as Fe(II) silicate in the mantle, and predominantly as Fe(III) oxides in the crust. From arguments of this kind, it appears that the detailed constitution of the Earth cannot be explained in terms of equilibrium condensation at any single temperature. There are, in fact, a number of good reasons why such a simple model would not be expected to work very well. Three problems which will be discussed briefly are:

(1) failure to reach chemical equilibrium in the condensation process;

(2) accretion of material which condensed over a range of temperatures;

(3) loss of volatiles by heating during or after accretion.

As far as (1) is concerned, the problem of heavy hydrocarbons, formed as non-equilibrium products in the reaction of CO and H_2, was mentioned in the previous section. At the higher condensation temperatures of most elements, there should be less difficulty in reaching equilibrium. This is hard to assess, however, without knowing more about the detailed conditions in the solar nebula, and especially its rate of cooling. The formation of planetary atmospheres (see later) suggests that a certain amount of volatile material—including a small fraction of noble gases—was trapped in the solid grains when they formed. It is likely that kinetic factors were more important at a later stage, after most of the Earth had accreted. Material forming the crust would then be well separated from that in the core, so that the different zones in the Earth would be prevented from coming into equilibrium with one another. This leads on to problem (2). It has been mentioned that the oxidation state of iron, in solids in equilibrium with the gas phase nebula, depends on the temperature. The structure of the Earth suggests that the material forming the inner parts may have condensed under hotter conditions than that in the outer layers. If there was a temperature gradient in the solar nebula, it is easy to understand how the planets can have collected solids which condensed at different temperatures. Although in the early stages of accretion most planetisimals probably had nearly circular orbits around the Sun, this was not the case later on. Near collisions between planetisimals would have deflected them into elliptical orbits, so that they could end up in a different temperature zone from the one where they originally condensed. Similar behaviour is still seen today with comets, which contain volatile material condensed in the outer solar system, but which occasionally come in close to the Sun. One can imagine, therefore, that planetisimals enriched in more volatile elements may have arrived on the Earth late in the accretion process, to give a surface layer containing, for example, water and fully oxidized Fe.

Turning now to problem (3), it is quite clear that all the planets have been subjected to some re-heating after they were formed. The major heat source in the Earth today comes from the decay of long-lived radioactive nuclides such as ^{40}K and ^{238}U. These were present in higher concentrations 4.6 billion years ago when the planets formed, and to their contribution was added that of shorter-lived species such as ^{26}Al. In addition to radioactive decay, there were two other important sources of energy during the accretion process. The first is the kinetic energy of impacting planetisimals. The other came from the separation of the phases of different density which make up the core and the mantle. Most theories suggest that metallic iron and silicate minerals condensed at very similar temperatures, and would have accreted together as a mixture. When other heat sources raised the temperature sufficiently for the iron to melt, it started to sink towards the Earth's centre. This liberated gravitational energy, which would provide further heat. In fact, this process can provide an embarassingly large amount of energy and, if it happened suddenly, it would heat the Earth sufficiently to boil off many of the more volatile elements. Some theories even suggest that the composition of the Earth owes more to the loss of volatiles by re-heating than it does to the original condensation process. According to this view, the inner planets formed from solids which condensed at quite a low temperature, perhaps 300 K, and which originally contained all except the extremely volatile elements in their solar system proportions. Volatile elements were then partially boiled off when the temperature rose due to radioactive heating. It is not easy to understand the composition of the Earth from this model, and it suffers from the same kind of difficulties as the simple picture of equilibrium condensation. For example, if the temperature were raised sufficiently to lose a high proportion of elements as the alkali metals, how could there by any water left? It is more likely that the re-heating processes happened gradually. Some boil-off of volatiles must have occurred, but probably not to the extent required in the low-temperature condensation theory. Indeed, one otherwise puzzling feature of the Earth can be explained if some volatile components were lost after the separation of the core and the mantle had occurred. Lithophile elements present in the outer layers would then be lost more readily than the siderophiles buried deep on the core. The relative abundances plotted in Fig. 4.6 do in fact show that of the elements in the '600–1200 K' range, the lithophiles are on average rather more depleted than the siderophiles.

Meteorites and comets

Meteorites, comets, and asteroids are small bodies in the solar system that have either failed to accrete into larger planets, or have come from the break-up of one or more planets. As solid bodies falling from space, meteorites provided scientists with their first direct extra-terrestrial chemical information. Indeed, it is no exaggeration to say that, in spite of recent space missions to the Moon and planets, the information derived from these insignificant pieces of rock, some only a few kilograms in mass, is still important in furthering our understanding of the early history of the solar system.

The diverse compositions of meteorites require a variety of origins. By far the greater number show signs of chemical differentiation by heating after they had accreted from the nebula. This, and other evidence, shows that they must have come from planets—probably located in the asteroid belt between the orbits of Mars and Jupiter—which broke up. In some cases the date of break-up appears to be quite recent on an astronomical time-scale, and was presumably caused by a collision between asteroids. Some of these meteorites are composed largely of magnesium silicates, and so are similar to the Earth's mantle. Others are predominantly metallic, and the iron–nickel mixture in these, with smaller amounts of other siderophilic elements, forms the usual basis for models of the Earth's core. However, it is another type of meteorite which has given most information about the formation of the solar system. These are the **carbonaceous chondrites**, already mentioned in Chapter 1 in the context of elemental abundances in the solar system. Figure 1.6 on p. 16 showed a comparison between the composition of the Sun and that of a typical carbonaceous chondrite. Apart from very volatile elements (H, C, N, O, and the noble gases), and ones rapidly consumed by nuclear reactions within the Sun (Li, Be, and B), the elemental abundances agree within experimental error. This strongly suggests that the meteorites have not been subject to appreciable reheating, and that they contain material remaining from the original accretion stage.

The overall composition of carbonaceous chrondrites is intermediate between that of the inner planets, dominated by silicates and metallic iron, and the outer ones, where the 'icy' fraction of extreme volatiles was also able to condense. They may well have originated from a region within the asteroid belt, where the temperature of the solar nebula was around 380 K, appropriate for condensation of all but the most volatile elements. However, they are not uniform in composition, and their name comes firstly from the higher proportion of carbon than is found in other types of meteorite, and secondly from the fact that (in common with many other meteorites) they contain many small mineral inclusions, known as **chondrules**. Both the chondrules and the dark-coloured carbonaceous matrix have been extensively studied.

Among the chondrules, the most interesting are those found to be rich in Al, Ca, and Ti. Their composition and mineralogy is very unusual, as they not only contain high proportions of these elements, but also of other rarer elements forming the early condensate fraction, predicted to solidify from the solar nebula at temperatures above 1400 K (see

Fig. 4.4). This class includes both lithophiles such as the lanthanide elements, and siderophiles such as Os, Ir, and Pt. It is thought that such high-temperature inclusions may represent the earliest material to have condensed from the nebula, although an alternative possibility—that they are *remnants* of solids which condensed at lower temperatures and then were partially re-volatilized—has also been considered. If they are, indeed, the first solids of the solar system, then they may have been formed close to the site of origin of the elements themselves, since as we have seen earlier the expanding gas cloud from a supernova may be the most favourable location for the nucleation of solid grains. In support of this idea is the unusual isotopic composition of these materials, which as explained in Chapter 6 suggests that they contain oxygen derived from different sources and not appreciably mixed since.

The composition of the high-temperature inclusions provides strong support for the theoretical condensation sequence described in previous sections. The carbonaceous material is remarkable in a different way. It is black, and composed of high-molecular-weight compounds quite different from the predictions of the equilibrium condensation model. As explained above, the gas-phase reactions of carbon compounds are very slow at temperatures below 500 K, where the equilibrium shifts from CO to CH_4 as the dominant form of carbon (reaction 4.8). Although a large amount of CH_4 was undoubtedly formed in the nebula, and is an important constituent of the outer planets, much carbon appears to have ended up in larger molecules. Carbonaceous meteorites contain not only heavy hydrocarbons, but also oxygen- and nitrogen-containing molecules, such as fatty acids, amino acids, and even the purine and pyrimidine bases which are important constituents of DNA. The discovery of such compounds in meteorites has fuelled speculations that life could have developed in space. Certainly, many of the necessary ingredients of life as we know it were present, but there is strong evidence that these ingredients themselves were produced by non-biological chemical reactions. The solid grains in the solar nebula contained compounds such as Fe_3O_4 that are good catalysts for the Fischer–Tropsch reaction between CO and H_2. Laboratory studies of this reaction show that it can produce the compounds found in meteorites, and that many features of the material found, such as the distribution of hydrocarbon isomers and the relative abundance of different amino acids, can be explained in this way. If living processes did not produce these compounds, it remains an open question as to whether their presence in meteorites (and possibly comets, see below) might have led to the development of life there.

Comets have been known since antiquity, and their appearance in past centuries has often occasioned a good deal of superstitious fear. The most celebrated is Halley's comet, named after the seventeenth-century scientist and colleague of Newton, who first showed that it returns periodically, every 76 years. The appearance in 1066 of what is now thought to be Halley's comet is represented in the Bayeux Tapestry, which commemorates the Norman conquest of England. Its return in 1986 provided the opportunity for close study by a number of unmanned space missions equipped with spectrometers and other instruments.

Like other comets, that of Halley is a small solid body travelling in a highly elliptical orbit around the Sun. In the outer part of the orbit it is invisible from the Earth. The often spectacular appearance of a comet (although Halley's was disappointing in 1986) arises when it comes closer to the Sun. The evaporation of material from the surface then produces a bright trail of gas and dust—the tail or **coma**—which is its most familiar feature. Spectroscopic measurements of comet tails show a variety of small molecules, radicals, and ions, dominated by the volatile elements H, C, N, and O. These and other observations have given rise to the 'dirty snow-ball' model, in which the nucleus of a comet is supposed to be made of ices of H_2O, NH_3, and CH_4, together with a smaller amount of rocky material. It is considered that comets are also remnants of the original planetisimals, but formed further out from the Sun than meteorites, in the region of the orbits of the giant planets. They may have avoided accretion into planets by being flung by the gravitational fields of Jupiter and Saturn into orbits far out in the solar system, beyond the most distant planets. In this theory, comets are only detectable after they have been perturbed, perhaps by the gravitational influence of a nearby star, into elliptical orbits which periodically bring them close to the Sun.

The species detectable in the tail of a comet result, at least partly, from the dissociation and ionization by high-energy (ultraviolet) radiation from the Sun. The central nucleus, and the immediate products of evaporation from it, are normally unobservable. One of the major aims of the missions to Halley's comet, especially of the European *Giotto* probe, was, therefore, to penetrate close to the nucleus, and to study the composition of gas and dust. Mass spectroscopy showed that, as expected, water was a dominant constituent, together with NH_3, CH_4, hydrocarbons, and CO_2. Large amounts of dust were found, containing the abundant elements H, C, N, O, Mg, Si, Ca, and Fe. The overall abundance of very volatile elements is greater than in carbonaceous chondrites, thus confirming the lower condensation temperature of cometary material. However, an unexpected feature was the extremely dark, almost black, appearance of the nucleus. Together with the detection of heavier hydrocarbons, this suggests the presence of carbon in the form of high-molecular weight compounds similar to those found in meteorites.

The evidence from comets and some meteorites—apparently 'primitive' bodies remaining from the origins of the solar system—seems generally to support the theoretical models of condensation used to interpret the chemical compositions of the planets. However, the occurrence of a wide variety of carbon compounds not predicted from equilibrium condensation does emphasize the limitations of this model. If so much carbon is present in the outer solar system in an involatile form, then it is possible that the inner planets, including the Earth, may contain more of this element than is normally thought. It has been suggested that large amounts of hydrocarbons could be trapped in the Earth's mantle, and even that the 'fossil' fuels normally thought to be derived from organic remains might come partly from this source. Many other uncertainties remain. It is worth mentioning, for example, the class of meteorites known as **enstatite chondrites**. The composition of these appears quite anomalous and suggests that they were formed in highly reducing conditions. They contain silicon in the metallic (mostly iron) phase and compounds such as CaS, TiN, and $Si_2N_2O_2$ which are not stable in the gaseous mixture normally assumed for the solar nebula, but which require a much higher [C]/[O] ratio. Possibly, these reduced compounds condensed in a different region of the nebula from most other meteorites, and they provide evidence that the composition may not have been at all uniform.

Chemical trends in the solar system

Table 4.2 summarizes some important physical properties of the planets. Much more information has become available in recent years, both as a result of remote sensing by Earth-bound or satellite-borne spectrometers, and from space missions to the planets themselves. These measurements have enabled the compositions of the outer parts of many planets—the atmospheres and solid surfaces—to be analysed in some detail. The best guide to the overall composition of the planets, however, still comes from estimates of the density.

The major trend in chemical composition is that which distinguishes the inner planets, Mercury, Venus, Earth (with the Moon), and Mars, from the outer ones. The inner planets have densities in the same range as that of the Earth, and are clearly similar in being largely composed of the 'rocky' fraction of elements that condense above 300 K. On the other hand, the outer planets are lower in density, and are dominated by the condensation of the 'icy' fraction of volatile hydrides. This is also the

Table 4.2
*Selected physical properties of the planets**

Planet	*Distance to sun* (10^8 *km*)	*Mass* (10^{24} *kg*)	*Radius* (10^3 *km*)	*Actual density* (10^3 *kg/m*^{3})	*Density corrected to 1 GPa* (10^3 *kg/m*^{3})
Mercury	0.58	0.33	2.44	5.44	5.3
Venus	1.08	4.87	6.05	5.27	3.9
Earth	1.50	5.98	6.38	5.52	4.0
(Moon	1.50	0.07	1.74	3.34	3.4)
Mars	2.28	0.65	3.39	3.96	3.9
Jupiter	7.78	1900	70	1.3	—
Saturn	14.3	570	60	0.7	—
Uranus	28.7	87	26	1.2	—
Neptune	45.0	100	25	1.7	—
Pluto	59	0.7(?)	3(??)	3(??)	

*From Smith (1979).

case for some of the satellites of Jupiter and Saturn. Their surfaces, photographed in the Voyager fly-by missions, are largely composed of a mixture of these ices. The major planets themselves all have atmospheres containing large amounts of hydrogen and helium: indeed, the overall compositions of Jupiter and Saturn are probably very close to that of the Sun itself. It is most likely that these planets formed from cores of rock and ice, which were sufficiently massive to pick up gaseous hydrogen and helium through their strong gravitational fields. In the interior of Jupiter and Saturn, the pressure is so high that hydrogen is thought to be present as a metallic liquid. Convection currents in this liquid give rise to magnetic fields rather similar to that which comes from the liquid Fe core of the Earth.

The major differences in composition between inner and outer planets can be easily understood in terms of a gradient of temperature in the solar nebula. The inner planets are formed from material which condensed above 300 K, whereas in the outer part of the solar system the temperature must have been low enough for 'ices' to condense as well. The differences between the inner planets themselves appear to be more subtle, however, and are still a matter of considerable controversy. As has been emphasized, even the Earth's composition is subject to many uncertainties, and it is clear that these must be greater in the case of the other planets. The data in Table 4.2 show that the Earth has the highest overall density, but an important correction needs to be applied before these values can be compared. The pressure at the Earth's centre is estimated to be 3.7×10^6 bar (3.7×10^{11} Pa), and this leads to a very considerable increase of density of the material there. For the smaller planets Mercury and Mars, the central pressures are probably around ten times less and the compression effect is smaller. In order to compare the densities, therefore, Table 4.2 shows values 'corrected' to a constant pressure of 10^4 bar. It must be emphasized that these values are somewhat uncertain, as both the estimates of central pressure, and the effect on the density, depend on assumptions about the internal composition. However, it is clear that the corrected density of Mercury is appreciably larger than that of the other planets, and that the value for the Moon is correspondingly smaller. It is difficult to tell whether the small differences between Venus, Earth, and Mars are really significant, but Mercury must have a relatively much larger core of reduced metal (presumably dominated by Fe) than the Earth, and the Moon much less, possibly none. The cause of these differences has been much disputed. In the case of Mercury the models of chemical condensation discussed previously suggest a number of possible reasons for its different composition.

1. The higher temperature in the inner part of the solar nebula may have led to the condensation of most Fe in the metallic form, with very little oxidation to Fe(II) or Fe(III).
2. If Mercury condensed at a sufficiently high **pressure**, the metallic phase of the 'metal + silicate' fraction might have condensed before the silicates.
3. There may have been **chemical inhomogeneities** in the nebula, resulting in a higher [C]/[O] abundance ratio in the inner parts; although it is not clear how this could have happened, it could lead to the condensation of metals before the major oxide fraction.
4. It has also been suggested that some unknown mechanism could have caused **segregation** of metallic and non-metallic solids in the nebula, with the Fe-rich metal concentrating closer to the Sun.

The importance of these and other factors in the formation of Mercury is uncertain. It is also unclear whether Mercury represents a 'special case', with some peculiar features due to its closeness to the Sun, or whether its different composition is part of a general trend in the inner planets. Most observations suggest that Venus and Mars are quite similar to the Earth in overall composition; differences appear in the surface rocks, but these are most likely the result of weathering effects in the very different atmospheres of the planets (see below).

The Moon occupies a special place in planetary investigations, partly because of the number of manned and unmanned space missions, which have enabled a fairly detailed picture of its surface composition to be built up. Detailed chemical analysis of surface rocks from a few sites has been supplemented by spectroscopic data from orbiting spacecraft, giving a more representative picture of the overall composition. The Moon is also unusually large relative to the size of the Earth compared with the satellites of other planets.

The major features of the chemistry of the Moon, compared with that of the Earth, are:

(1) a much lower abundance of **iron**;

(2) a very much lower concentration of **volatile** elements (even those, such as alkali metals, in the >600 K range are depleted, and for elements volatile below 600 K this depletion is very striking;

(3) a higher concentration of elements in the **early condensate** fraction, including Al, Ca, Ti, and U.

As with the planet Mercury, there exists at the present time no general consensus on how these differences are best explained. Indeed, none of the simple theories proposed for the Moon's origin can explain its composition at all easily. One theory is that the Earth and Moon formed at the same time, by **simultaneous accretion** of planetisimals in similar orbits. This seems to be ruled out, as the material making up both bodies should be of very similar composition. The other traditional theories are (a) that the Moon formed from the Earth by **fission**, and (b) that the Moon initially accreted as an independent planet, and was subsequently **captured** by the Earth. Neither of these models is very satisfactory as it stands. The low abundance of iron (and correspondingly high silicate content) could be understood from the fission model if the Moon originated from part of the Earth's mantle, but it is less easy to see how the enrichment in early condensate and the loss of volatiles can have come about. On the other hand, calculations on the capture theory show that an in-coming planet would almost certainly break up into small pieces in such an encounter. A picture of **disintegrative capture**, followed by reconsolidation of the fragments is, therefore, more realistic. The best model may be one which involves some compromise between the different simple theories. One can imagine a near collision late in the accretion stage, leading to a break-up of both the in-coming body and part of the Earth. The energy liberated in such a collision might have been sufficient to boil off a large proportion of the volatile elements from the Moon before it reformed. As suggested in the previous section, it is also likely that the Earth acquired some volatile components rather late in the accretion process. The smaller Moon has a weaker gravitational field and, especially if it was still in many pieces, it would have been much less efficient at capturing these 'late volatiles'.

As is clear from this discussion, there are still many uncertainties in the formation of the planets and the Moon in particular. The chemical evidence that has become available in recent years has enabled many of the 'wilder' speculations to be eliminated, but above all, it has sharpened the realization of planetary scientists that the origin of the solar system was a very complex process.

The atmospheres of the planets

The atmospheres of the planets are generally more accessible to remote study than are the solid surfaces and interiors. Through measurements of infra-red and other spectra, quite detailed information about the chemical composition has been obtained. Although the atmospheres make up a very small fraction of the total mass of most planets (only 10^{-4} per cent for the Earth, for example), their chemistry is, nevertheless, very interesting, and can give many clues to the origin and nature of the planets as a whole. The Earth's atmosphere is also essential for the support of life, and theories of the origin of life rely heavily on ideas about its original composition and how this has evolved. This itself, and the possibility of finding life on other planets, forms one strong motivation for looking at the atmospheres of the planets.

Table 4.3 shows some data on planetary atmospheres, summarizing the major features of their composition and surface pressure. The atmosphere of Jupiter is fairly typical of the other large outer planets. Mercury, the Moon, and other smaller bodies in the solar system have no significant atmosphere. However an atmosphere originates, one crucial requirement is that the planet's gravitational field is strong enough to retain molecules in the gas phase. This is determined by the **escape velocity**

$$\mathrm{V_e} = (2GM/R)^{1/2} \tag{4.9}$$

where M is the mass and R the radius of the planet, and G the universal constant of gravitation. Values of escape velocity are given in Table 4.3. In the upper atmosphere, where the density is so low that atoms and molecules can travel a long distance without collisions, they will escape the gravitational pull of the planet if their kinetic velocities exceed V_e. The probability of this can be estimated from Maxwell–Boltzmann thermal distribution of velocities, and depends strongly on the mass of the atom or molecule as well as on the temperature. On Earth, for example, the relevant part of the upper atmosphere has a rather high temperature, about 600 K. At this temperature, the fraction of H atoms with thermal kinetic velocities exceeding 11 km/s (V_e for the Earth) is

Table 4.3
*Planetary atmospheres**

Planet	*Escape velocity (km/s)*	*Surface temperature (K)*	*Surface pressure (bar)*	*Principal constituents (per cent by volume)*
Venus	10.3	732	90	CO_2 (96.5), N_2 (3.5), SO_2 (0.015)
Earth	11.2	288	1	N_2 (78.1), O_2 (20.9), H_2O (variable), Ar (0.9), CO_2 (0.03)
Mars	5.0	223	0.006	CO_2 (95.3), N_2 (2.7), Ar (1.6), O_2 (<0.1), H_2O (0.03)
Jupiter	60	(no true surface)		H_2 (80), He (20)
Titan	2.1	95	1.6	N_2 (82), Ar (12), CH_4 (3), H_2 (0.2)

*From Wayne (1985).

10^{-6}. This is high enough for hydrogen to be lost rapidly from the Earth's atmosphere. On the other hand, only one in 10^{84} of O atoms will exceed the escape velocity, so that the escape of oxygen is negligible. Escape velocities for Mercury and the Moon are 4 and 2 km/s, respectively. These values are too low for a planet to retain an appreciable atmosphere, especially at the high temperature prevailing on the surface of Mercury. On the other hand, the giant planets Jupiter an Saturn, with their high V_e values and low temperatures, have atmospheres dominated by the light species H_2 and He, which are not retained by the inner planets.

How then did atmospheres originate? There are basically four possible mechanisms.

1. Capture of gases directly from the solar nebula, before it dispersed.
2. Accretion on the planet's surface of solid material, containing volatile constituents which condensed in the outer solar system.
3. Capture of atoms from the 'wind' of particles blown off from the Sun.
4. Outgassing during the reheating phase, of volatile material contained in the planetisimals from which the planet accreted.

In the case of the large outer planets, mechanism 1—capture directly from the gaseous nebula—undoubtedly accounts for most of their atmospheres. The composition of these is close to that of the original nebula and no other mechanism can explain the enormous quantity of gas involved. The large gravitational pull of the heavy planets was clearly able to attract the gases before the nebula dispersed. The inner planets, however, were either too hot or too small to retain an atmosphere while there was sufficient gas around to be picked up. The strongest evidence that their atmospheres were not formed in this way comes from the abundance of noble gases. Apart from ^{40}Ar, coming from the radioactive decay of ^{40}K since the formation of the planets, all these elements have an extremely low relative abundance. For example, in the solar system Ne is almost as abundant as N, but the ratio of concentrations in the Earth's atmosphere is only 10^{-5}. Since Ne and NH_3 (the dominant form of N in the nebula) have similar masses, they should have been acquired equally easily from the gas phase, and it is impossible on this basis to explain the discrepancy in abundance on Earth.

Mechanism 2—accretion of volatile material from solid bodies after the formation of the planets—is a little harder to dismiss. Sometimes this is called the 'cometary hypothesis', as it is envisaged that the planets were bombarded by comet-like bodies containing volatile elements. It is certain that *some* volatile material must have been acquired in this way, but most calculations suggest that it would not be nearly sufficient to account for the Earth's atmosphere. (It has also been suggested that life may have arrived on Earth in a similar way, in some kind of primitive organism contained in comets; this idea is also difficult to dismiss, although it is not widely believed.) The acquisition of solar wind—mechanism 3—also cannot give nearly enough gases, but it is possible that a proportion of rare elements such as noble gases might have been collected from this source.

Most theories of the formation of the atmospheres of Venus, Earth, and Mars are based on mechanism 4. It is supposed that small amounts of volatile compounds were included in the solids from which the planets accreted. Water can certainly be taken up into silicates, to form clay minerals (see Chapter 5). Nitrogen and carbon could be present in various forms: for example, as nitrides and carbides of metals such as Fe, as carbonates or nitrates, and also as their hydrides which can form **clathrate compounds**, in which molecules are included in other crystal lattices. Even small amounts of the unreactive noble gases might be trapped in this way in the grains condensing from the solar nebula. After the planets had accreted, radioactive decay and other energy sources heated them, and this would have led to the **out-gassing** of the volatile constituents. Volcanic gases today contain CO_2, N_2, water vapour, and many minor components such as H_2S, SO_2, and HCl coming from the decomposition of minerals deep in the Earth. In the early stages of re-heating, such volcanic activity was probably much more violent, and would have contributed large amounts of these gases to the early atmosphere. The more reactive species such as SO_2, H_2S, and HCl would have reacted quite quickly with surface rocks, leaving an atmosphere dominated by N_2, CO_2, and water vapour. It has often been suggested—especially in the context of theories of the origin of life—that the early atmosphere might also have contained large amounts of CH_4 and NH_3. Although these gases are not common in volcanic emissions today, they might well have been produced in the earlier stages of degassing. Indeed, there are people who believe that a great deal of methane might still be trapped in the mantle. However, such strongly reducing components would probably not have survived for long in the atmosphere. One species that was certainly not present in significant proportions in the early atmosphere was free molecular oxygen. The origin of this is considered later.

The atmospheres of Venus and Mars have clearly evolved in a very different way from that of the Earth, but this is entirely consistent with the out-gassing hypothesis. The smaller planet, Mars, was probably subject to less out-gassing, but Venus is very similar to the Earth in size and overall composition, and it is likely that the early atmospheres of these two planets were very similar. It is not certain whether there was ever liquid water on Venus, but even if there was, the proximity of the Sun would lead to a higher surface temperature. This would give a larger pressure of water vapour in the atmosphere, which would increase the heating, by strongly absorbing infra-red radiation re-emitted from the planet's surface. Greater heating would, in turn, give more atmospheric water vapour, so leading to a **runaway greenhouse effect**, with a surface temperature sufficient to vaporize all the water. There is very little water in the present Venusian atmosphere, and this is due to photochemical processes in the upper atmosphere, initiated by hard UV radiation from the Sun. The first step is that of **photodissociation** of water:

$$H_2O + h\nu = H + OH \tag{4.10}$$

where $h\nu$ represents in the energy of a high-energy photon. The hydrogen rapidly escapes into space, and the OH radicals undergo a series of reactions leading to the production of O_2. This in turn oxidizes surface rocks and is present in only very low concentrations on Venus. The idea that a large quantity of water has been lost in this way is supported by the observations of the isotopic abundance. Deuterium forms about 1 per cent of the hydrogen on Venus, compared with only 0.015 per cent on Earth. This can be explained by the differential loss of the lighter isotope, expected both on the grounds of chemical reactivity and gas kinetic effects (see Chapter 6). The other differences in the atmosphere of Venus can also be readily accounted for. The amount of N_2 in the atmosphere is not very dissimilar to that on Earth. Even the dramatic difference in the CO_2 partial pressure (0.03 bar on Earth, compared with nearly 90 bar on Venus) can be explained by the fact that CO_2 on Earth is mostly combined chemically in sedimentary rocks, in the form of carbonates. Thus, the *total* amount of CO_2 is again comparable on the two planets; presumably, carbonates are not formed on Venus due to the acid atmosphere and the high temperature. It is the lack of water which probably accounts for the presence of acid gases such as SO_2 and HCl. The thick clouds which cover the surface of Venus are composed largely of sulphuric acid, formed by the oxidation of SO_2. Although H_2SO_4 is present on the Earth (nowadays partly as the result of burning sulphur-containing fossil fuels) it is rapidly precipitated by rainfall and reacts with surface rocks.

Mars has a much thinner atmosphere than the Earth. This must be partly due to the lesser degree of out-gassing expected for a smaller planet. The escape velocity is also much lower, however, and a fair proportion of the out-gassed material must have been lost into space. Evidence again comes from isotopic abundances. In this case the $^{15}N/^{14}N$ abundance ratio measured in the Martian atmosphere is significantly larger than that on Earth. As with D and H the heavier isotope is less easily lost, but the efficiency of fractionation of nitrogen isotopes is quite low. It can be deduced that a high proportion of the original N_2 must have been lost to account for the observed enrichment in ^{15}N. There is quite a large amount of water present on Mars, and some surface features suggest that some of this may have been liquid in the past. The present temperature is so low, however, that most water is now present as ice, with very little in the liquid or vapour phase.

The Earth has escaped both the 'runaway greenhouse effect' of Venus, and the reverse 'runaway freeze-out' which may have affected Mars. It is unique in the solar system in having most of its surface in the temperature range where water is liquid. The low temperature of the mid-atmosphere also prevents much water vapour from reaching the upper atmosphere and undergoing photodissociation. The other unique feature of the Earth's atmosphere is the presence of molecular oxygen. This was almost certainly not present originally. A little oxygen must have been produced by photodissociation of water, but this would have been used up rapidily in the oxidation of surface rocks. That the concentration of oxygen in the primitive atmosphere was very low is suggested by a great deal of geological evidence, especially the existence of sedimentary rocks laid down 3–4 billion years ago containing elements in reduced forms such as Fe^{2+}, unstable in the presence of free oxygen. A major geological event of the period between 2 and 3 billion years ago was the formation of so-called **banded ironstones**, laid down by the oxidation of iron from the water-soluble Fe^{2+} form to the insoluble Fe^{3+} state. The oxidation of iron and other elements, such as uranium, shows that the O_2 content of the atmosphere was starting to rise before 2 billion years ago. This must have been due to the development of biological systems which build up organic molecules from CO_2 by **photosynthesis**, liberating O_2. Thus, Earth owes its unique atmosphere to the presence of life; in return, however, the present life-forms owe much to the atmosphere itself. Not only did the presence of O_2 lead to the evolution of air-breathing organisms, but it was essential if life was to move out of the ocean onto dry land. This is because UV radiation is very damaging to biological systems. Only with the increase of O_2 content in the atmosphere could an ozone layer develop, and prevent this radiation reaching the Earth's surface.

Most of the outer planets, as mentioned previously, have massive atmospheres largely dominated by hydrogen and helium. That of Jupiter, shown in Table 4.3, is fairly typical. Spectroscopic measurements show many other volatile species, especially hydrides such as NH_3, CH_4, H_2S., and PH_3. These undergo a variety of poorly understood photochemical reactions, giving some strongly coloured species which account for the appearance of these planets. The other data given in Table 4.3 are for Titan, the large satellite of Saturn. It is interesting as it is the only satellite in the solar system large enough to support an appreciable atmosphere. As Table 4.3 shows, this is dominated by N_2, Ar, and CH_4, but again, many minor species such as larger hydrocarbons are present and may be photochemically produced. It is likely that nitrogen was originally present largely as NH_3; this has undergone photodissociation in a way similar to the water on Venus and been left as molecular nitrogen.

Summary

As well as the luminous matter in stars, dark clouds of interstellar gas and dust also contain important reservoirs of elements in space, from which new stars and planetary systems can form. Many molecules, dominated by the elements H, C, N, and O, have been detected in such clouds, and there is also evidence for solid grains containing carbon and less volatile elements, probably in the form of metals and silicates.

Much evidence suggests that the Earth and other planets formed at the same time as the Sun, 4.6 billion years ago. Theories indicate that grains of solid material accreted by gravitational forces, first into small bodies known as planetisimals and then, more slowly, into larger planets. Comets and some types of meteorite are probably remains of the original planetisimals and can, therefore, give information about the chemical composition of the material from which the planets were made. Calculations on the condensation of solids from the gas phase show a wide temperature range. The composition of the Earth and other inner planets is dominated by elements which condensed above about 600 K, whereas the temperature in the outer solar system was low enough for formation of ices of HO, NH_3, and CH_4. However, many difficulties still remain in interpreting the detailed chemistry of the planets, and especially of the Moon which is highly deficient relative to the Earth in moderately volatile elements.

Small amounts of volatile elements were trapped in the solid grains during the accretion process. Out-gassing of these components, due largely to radioactive heating, gave rise to the atmospheres of the inner planets. The Earth's atmosphere has evolved in a unique way: firstly, because the surface temperature is in the range where most water is present in liquid form; and secondly, because the development of photosynthesis by living organisms has given a large concentration of uncombined oxygen.

Further reading

The books by Duley and Williams (1984), and Dyson and Williams (1980) review the chemistry and the physics of interstellar clouds. That by Suess (1987) gives a useful, but rather limited account of cosmochemistry in relation to the solar system.

Accounts of the origin of the Earth and the planets can be found in many books, including McElhinney (1976) and Ringwood (1979). The latter is quite detailed and useful, but puts forward a model for the formation of the Moon, in particular, which is not widely accepted. Indeed, the non-specialist reader will find that the detailed literature in this area tends to be contradictory and highly confusing. A review especially used in writing this chapter is that of Smith (1979); the important cosmochemical work by Grossman and Larimer (1974), and by Ganapathy and Anders (1974) must also be mentioned.

Interesting discussions of carbonaceous chondrites, and their relevance to the early solar system, are given by Anders *et al.* (1973), Mason (1975), and Wilkenig (1975). Spinrad (1987) reviews recent work on comets. Wayne (1983) discusses the chemistry of planetary atmospheres, including their origin and evolution.

5
The distribution of elements on the Earth's surface

Nearly all the elements accessible to us are obtained from the thin crust which forms the outer layer of the Earth. For this reason, the chemical composition of the crust has an importance which far outweighs its relatively insignificant mass—about 0.5 per cent of the Earth's total. This is the subject of this chapter. It continues the story begun in the previous chapter, which showed how the composition of the Earth as a whole can be understood from the processes of chemical fractionation which accompanied the formation of the solar system. The major differentiation of elements within the Earth—forming a metallic core and the outer silicate layers—almost certainly took place during this formation stage. Further chemical fractionation has continued over a much longer time scale, however, in the gradual evolution of the Earth's crust and oceans. One of the major developments in Earth sciences in recent decades has been the realization that this evolution is continuing today, and the modern theory of **plate tectonics**—described briefly in a later section—now provides a more satisfactory basis for understanding the chemical composition of the Earth's surface.

The composition of the crust

The composition of the Earth's crust was shown in Fig. 1.2 on p. 9. Figure 5.1 shows how the abundances of elements in the crust compare with those of the Earth as a whole. The ratios of crustal to whole Earth abundances are plotted on a logarithmic scale, against the position of elements in the periodic table. The idea behind this plot is the same as that used in Chapter 4 to show the composition of the Earth relative to the solar system as a whole. In the present case, Fig. 5.1 shows the chemical fractionation of elements that has accompanied the formation of the crust. It is easy to see that some elements, high up in Fig. 5.1, tend to be concentrated in the crust relative to the bulk of the Earth, whereas other elements are quite strongly depleted in the crust.

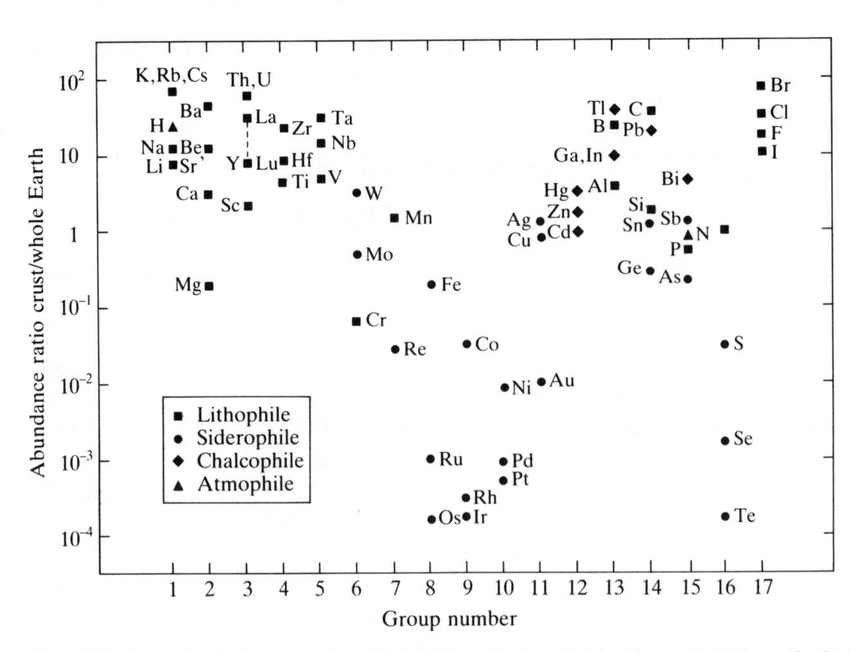

Fig. 5.1. Comparison of crustal with whole-Earth abundances of the elements. Elements high up in this plot are enriched in the crust relative to the Earth as a whole; ones low down are depleted and concentrated in the inner parts of the Earth.

The geochemical classification explained previously is used in Fig. 5.1 to show each element and certain trends are immediately apparent. Lithophile elements—which occur as oxides and other ionic compounds—are relatively abundant in the crust, whereas the siderophiles—entering the metallic phase of the core—are low in relative abundance. However, this correlation is by no means perfect and some elements seem to be rather anomalous. For example, the strong lithophiles magnesium and chromium are surprisingly low in Fig. 5.1, and are less abundant in the crust than in the Earth as a whole. On the other hand, some elements classed as siderophiles have a comparatively high concentration in the crust. Some of these anomalous elements (for example Mo and W) are borderline in their geochemical behaviour, but it is also important to remember that no less than two-thirds of the Earth's mass is made up of the **mantle** (see Fig. 1.3 and Table 1.2). This is also composed of silicate and other oxide minerals, and so contains the major portion of many lithophile elements. The low position of magnesium and chromium in Fig. 5.1 comes about because they tend to concentrate in the mantle, where their abundances are more than ten times their values in the crust. On the other hand, the high crustal abundance of many lithophile elements is a reflection of much lower concentrations in the mantle, by a factor of ten for example, for the alkali metals. Thus, the composition of the crust is determined not only by the lithophile/siderophile division, but also by the 'upward mobility' of some elements relative to others. The geological and chemical factors controlling this will be explained in the following sections.

Although the overall composition of the crust is important, it can obscure an equally significant feature, the remarkable lack of chemical uniformity. Some elements are actually rather evenly distributed throughout the surface of the Earth, but many others occur in high concentrations in specific minerals, and sometimes these minerals themselves are found predominantly in certain parts of the world. Such non-uniformity of distribution is clearly of great economic importance. For example, the platinum group of metals (Ru, Rh, Pd, Os, Ir, and Pt) are strong siderophiles, and their overall crustal abundance is so low that it is hard to estimate. Some of these elements have very desirable catalytic properties, and their exploitation depends on the fact that they can be found in metallic form and ores where their concentration may be more than a million times that of the crustal average. Ores with concentration factors of hundreds or thousands are also necessary for the commercial exploitation of many important metals, such as copper, tin, lead, and uranium.

Table 5.1 gives a summary of the main mineral and other sources of the elements. The diverse geological origins of the ores, noted in the final column, will be explained later. It is important to note that the main source of each element depends on economic, as well as on geological and chemical factors. Thus, the elements Al, Si, and Fe are all extremely widespread in the crust, but the economically important sources represent particularly concentrated forms, from which other elements have been largely separated by geochemical processes. In the case of elements used in small quantities, on the other hand, different factors may operate. For example, As occurs in concentrated form in the mineral arsenopyrite, FeAsS. Nevertheless, it is widespread in a large number of sulphide ores, such as those of Cu and Pb, and is most easily obtained as a byproduct of the processing of these. Inspection of Table 5.1 will show that several other elements of lesser abundance and of relatively minor use, come from similar sources.

Although Table 5.1 gives only a limited picture, it does illustrate the wide diversity of mineral types. One particular observation which can be made is the way in which elements—either of similar or different chemical behaviour—tend to be associated together. The pattern of element affiliations in minerals is a highly complex subject, which it would be quite impossible to do justice to in a limited space. Nevertheless, it is interesting to look briefly at the summary shown in Fig. 5.2. Each symbol in this figure connects a pair of elements which are

Table 5.1
Principal sources of elements

	Mineral or other source	*Formula*	*Origin*
H	natural gas	hydrocarbons	biological
He	natural gas		radioactive decay of Th, U
Li	spodumene	$LiAlSi_2O_6$	pegmatites
Be	beryl	$Be_3Al_2Si_6O_{12}$	pegmatites and placer deposits
B	borax and other Na borates		volcanic regions (hot springs)
C	diamond		kimberlies, and placer deposits
	graphite		metamorphic
N	atmospheric		
O	atmospheric		
F	fluorite	CaF_2	widespread
Ne	atmospheric		
Na	rocksalt	NaCl	evaporites
Mg	dolomite	$CaMg(CO_3)_2$	sedimentary
Al	boehmite	AlO(OH)	weathering product of aluminosilicates
	gibbsite	$Al(OH)_3$	
Si	quartz	SiO_2	widespread
P	apatite	$Ca_5(PO_4)_3(F,Cl)$	widespread
S	native	S	bacterial, volcanic
Cl	rocksalt	NaCl	evaporite
Ar	atmospheric		radioactive decay of ^{40}K
K	saline waters (brines)	KCl	
Ca	calcite	$CaCO_3$	sedimentary (marine organisms)
Sc	uranium ores		
Ti	ilmenite	$FeTiO_3$	weathering (resistates)
	rutile	TiO_2	

V	carnotite	$K(UO_2)(VO_4)1.5H_2O$	sedimentary
Cr	chromite	$FeCr_2O_4$	magmatic
Mn	braunite	$3Mn_2O_3 \cdot MnSiO_3$	weathering
Fe	haematite	Fe_2O_3	weathering (oxidate)
Co	carrollite	$Cu(Co,Ni)_2S_4$	magmatic
Ni	pentlandite	$(Fe,Ni)_9S_8$	magmatic
Cu	chalcopyrite	$CuFeS_2$	hydrothermal
Zn	zinc blende (sphalerite)	ZnS	hydrothermal
Ga	aluminium ores		
Ge	zinc and other non-ferrous ores		
As	copper and lead ores		
Se	copper and other non-ferrous sulphide ores		
Br	brines	Br^-	
Kr	atmospheric		
Rb	lithium minerals		
Sr	strontianite	$SrSO_4$	various
Y	lanthanide minerals		
Zr	zircon	$ZrSiO_4$	weathering (resistate)
	baddeleyite	ZrO_2	
Nb	columbite	$(Fe,Mn)Nb_2O_6$	pegmatites
Mo	molybdenite	MoS_2	pegmatites, hydrothermal
Tc	nuclear reactors		fission of ^{235}U
Ru	native		igneous sulphide ores and placer deposits
Rh			
Pd			
Ag	copper and other non-ferrous sulphide ores		
Cd	zinc ores		
In	zinc and lead ores		
Sn	cassiterite	SnO_2	hydrothermal, placer deposits
Sb	stibnite	Sb_2S_3	hydrothermal

Table 5.1 (*cont.*)

	Mineral or other source	*Formula*	*Origin*
Te	copper and other non-ferrous sulphide ores		
I	brines		
Xe	atmospheric (also fission products)		
Cs	mostly Li minerals		
Ba	barite	$BaSO_4$	various
La-Lu	monazite	$LnPO_4$	weathering (resistate)
	bastnaesite	$Ln(CO_3)F$	
Hf	zirconium minerals		
Ta	tantalite	$(Fe,Mn)Ta_2O_6$	pegmatite
W	scheelite	$CaWO_4$	hydrothermal, pegmatite
	wolframite	$(Fe,Mn)WO_4$	hydrothermal
Re	molybdenum ores		
Os, Ir, Pt	native		sulphide ores and placer deposits
Au	native		hydrothermal, placer deposits
Hg	cinnabar	HgS	volcanic, hydrothermal
Tl	zinc and lead ores		
Pb	galena	PbS	hydrothermal
Bi	non-ferrous sulphide ores		
Ra	uranium ores		radioactive decay of ^{238}U
Th	lanthanide minerals		
U	uranitite	UO_2–U_3O_8	various
Pu	nuclear reactors		neutron bombardment of ^{238}U

Sources: Greenwood and Earnshaw (1984).

Fig. 5.2. Element affiliations in minerals of the Earth's crust. Each symbol marks a pair of elements that are frequently found together. The different symbols show: ●, elements found directly bonded together; ○, indirect associations, for example between lithophilic metals and silicon; +, elements that are often exchangeable at the same lattice site. Smaller symbols indicate less common associations.

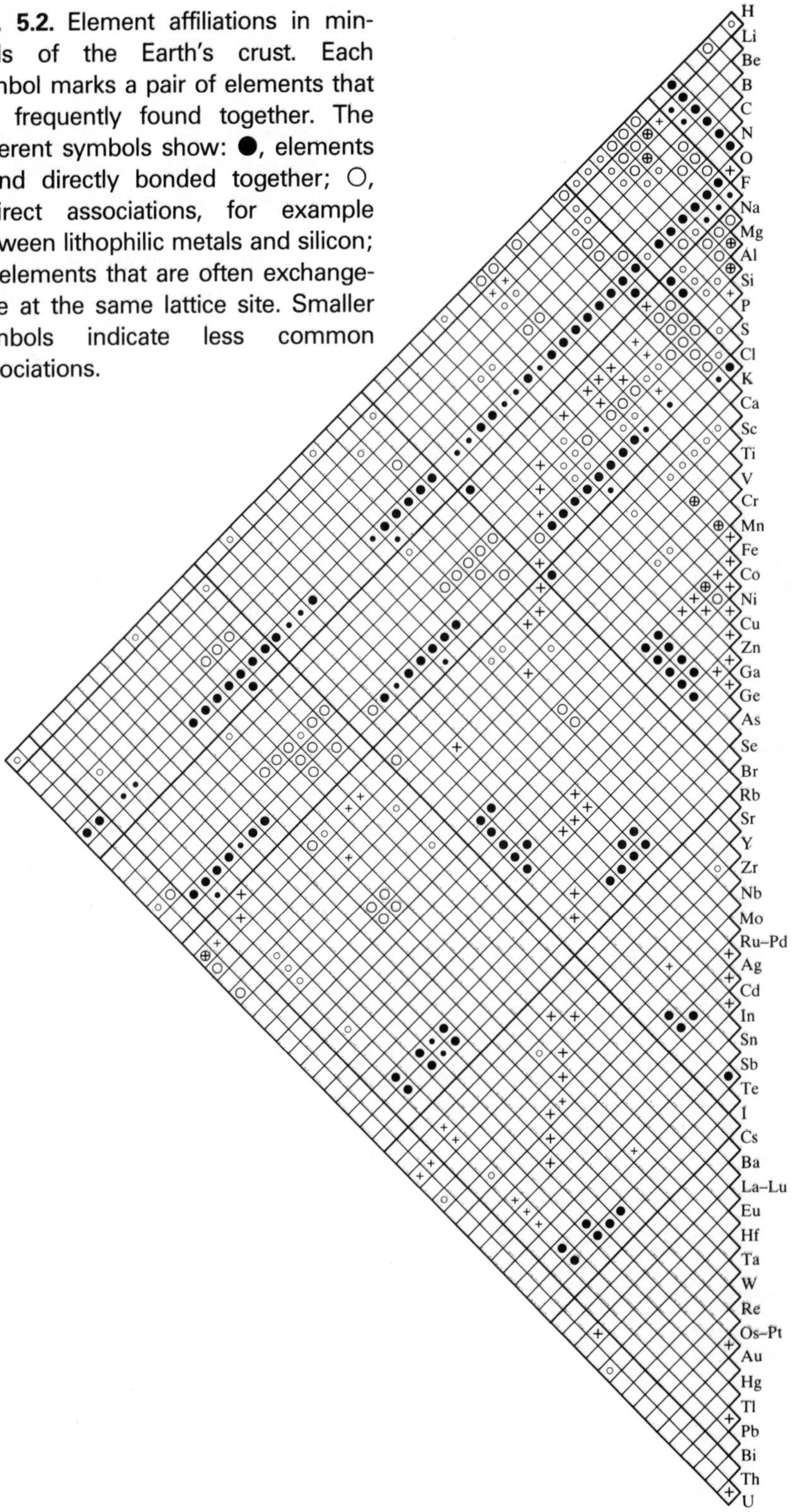

H
Li
Be
B
C
N
O
F
Na
Mg
Al
Si
P
S
Cl
K
Ca
Sc
Ti
V
Cr
Mn
Fe
Co
Ni
Cu
Zn
Ga
Ge
As
Se
Br
Rb
Sr
Y
Zr
Nb
Mo
Ru–Pd
Ag
Cd
In
Sn
Sb
Te
I
Cs
Ba
La–Lu
Eu
Hf
Ta
W
Re
Os–Pt
Au
Hg
Tl
Pb
Bi
Th
U

commonly associated in the same mineral. (Some less common associations are marked with smaller symbols). The three symbols are used to show different ways in which the affiliation can arise.

1. Solid circles connect pairs of elements that are found directly bonded together; that is, at neighbouring sites in the crystal lattice. Thus, lithophilic elements are found bonded to oxygen (and less commonly to the halogens F, Cl, etc.), chalcophilic elements to sulphur, and less commonly to As, Se, and Te.

2. Open circles represent indirect associations arising from the elements occupying different types of (non-adjacent) lattice site. Commonly, these may arise when the two elements concerned are bonded to a common atom. For example, B, C, P, and Si are generally associated with lithophilic metals in this way, in borates, carbonates, phosphates, and silicates, respectively.

3. Another form of association which is responsible for much of the complexity of mineral chemistry is when two or more elements may replace each other in varying proportions in the same lattice sites. These are marked by a cross in Fig. 5.2. Elements associated in this way are often in neighbouring positions (either horizontally or vertically) in the periodic table and, clearly, the possibility of this kind of replacement depends on having a similar chemical behaviour. In fact, the *size* of the lattice site involved is generally the controlling factor and many of the associations marked can be understood on this basis, using the **ionic radii** for elements, discussed briefly in Chapter 2 (see Fig. 2.9 on p. 47). The first-row transition metals Cr, Mn, Fe, and Ni commonly replace Mg in this way. It is also generally easier to replace Na by Ca, which is of similar size, than by K which is appreciably larger. In cases like this the different charge on the ions (Na^{+} as opposed to Ca^{2+}) must be compensated by some other replacement, for example, of Si by Al. Thus, we obtain the common mineral plagioclase feldspar, which can be formulated (Na,Ca) $(Al,Si)_4O_8$: the notation (Na,Ca), for example, shows that the two elements may be present in varying proportions. Essentially all minerals, even those formulated as 'pure' stoichiometric compounds, show some degree of replacement of this kind, and in many of the commoner silicate and sulphide minerals which make up the crust ten or more elements may be present in proportions ranging down to 0.1 per cent.

The formation of the crust

Geologists classify rocks into three major types, depending on their origin.

1. **Igneous rocks** are formed from the solidification of molten material—**magma**—coming from high-temperature regions deep in the Earth.
2. **Sedimentary rocks** arise from the action of air, water, and biological processes at the Earth's surface, on minerals of igneous or other origin.
3. **Metamorphic rocks** are formed by the modification of igneous and sedimentary rocks by high temperatures and pressures, but without the melting process involved in igneous rocks.

A common igneous rock of the continental crust is **granite**; **limestone, sandstone,** and **clay** are typical sedimental rocks; and examples of metamorphic rocks are **marble** and **slate**.

It is the magmatic processes involved in forming igneous rocks that lead to the major differentiation of the crust from the underlying mantle. Both these processes and those giving rise to sedimentary rocks depend to a large extent on liquid–solid equilibria: in the case of igneous rocks such equilibria involve molten silicates and other rocks, whereas water is the important liquid phase in sedimentary processes. As well as forming new rocks, therefore, both these phenomena can lead to major redistributions of elements, by virtue of differences in solubility in the corresponding liquid phase. The formation of metamorphic rocks, on the other hand, largely involves reactions between solid phases, and so the changes in element distribution which result from these processes are less important.

The development of the theory of **plate tectonics** has in recent years led to a much better understanding of the way in which crustal rocks originate from the mantle. According to this theory, parts of the Earth's crust are being continually formed and destroyed in cyclic processes resulting from slow convection currents in the mantle. The primary driving force for such convection comes from heat sources within the Earth, the most important being the energy released by the radioactive decay of long-lived nuclides. At the present time the major contribution comes from the isotopes ^{232}Th, ^{238}U, and ^{40}K, but in the early stages of the earth's history, there were probably other radioactive species present, such as ^{26}Al and ^{244}Pu, which have decayed much more rapidly. Figure 5.3 shows an estimate of the heat production from these different isotopes as a function of time. There was a rapid decline during the first 10^8 years or so, as the shorter-lived nuclides decayed, and then a rather slower decline towards the present value.

The rocks in the mantle are poor conductors of heat and, in the absence of convection, the energy generated by radioactive decay would soon raise the temperature to the melting point. In fact, the mantle is *not* liquid in the normal sense. Seismic studies, measuring the propagation of

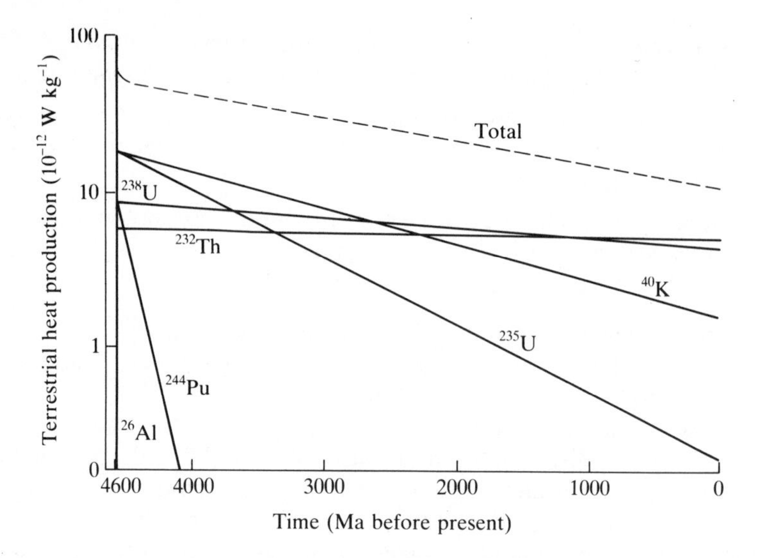

Fig. 5.3. Heat production from the decay of radioactive species in the Earth, expressed as a global average per unit mass, showing the fall since the origin of the Earth. (From Brown and Musset 1981.)

shock waves from earthquakes, show quite clearly that the mantle is resistant to rapid deformation and displays the mechanical properties characteristic of solid material. In a similar way to many solid polymers and even silicate glass, however, it seems that the mantle can slowly deform or **creep**, so that on a geological time-scale it behaves more like a very viscous liquid than a solid. Thus, slow convection currents can be set up, which are not only important in forming the crust, but can also move pieces of crust around on the Earth's surface over periods of hundreds of millions of years. One of the great achievements of the plate tectonic theory is that it can account for the phenomenon of **continental drift.** It is now well established that around 200–300 million years ago the major land-masses of the Earth were joined in a 'super-continent' known as **Pangea.** The Atlantic Ocean started to form 100 million years ago and is still widening slowly. Prior to about 500 million years ago it is thought that North America and Europe were separated by the **Iapetus** Ocean. This closed up, causing the continents to collide and, thus, giving rise to the mountain ranges of Scandinavia, north western Britain, and the Appalachians in the eastern United States. More recent (and apparently more violent) continental collisions have produced The Alps and the Himalayas.

The basic tectonic processes forming the crust are shown (in a very simplified form) in Fig. 5.4. Upwelling convection currents bring mantle

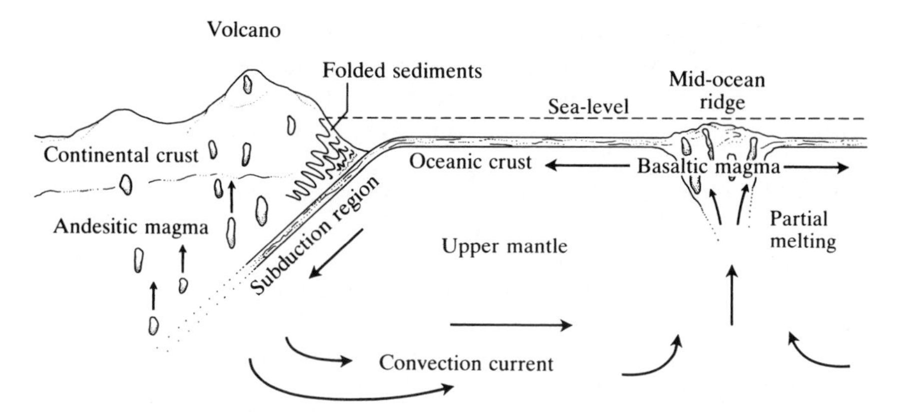

Fig. 5.4. Simplified illustration of the tectonic processes, driven by mantle convection, that form the oceanic and continental crust.

rocks towards the surface, mostly at those places on the Earth which form **mid-ocean ridges**; one such ridge, for example, runs mid-way down the entire length of the Atlantic Ocean. As rocks come close to the surface the pressure decreases, and a point is reached at which partial melting occurs. Liquid magma rises, occasionally welling out of the surface in a volcano. More frequently, it cools and solidifies below the surface, producing the type of rock known as **basalt**, which forms the major part of the oceanic crust. This crust is being formed continually along mid-ocean ridges, causing an outward motion of the surrounding sea floor at a rate of a few centimetres per year.

In other parts of the Earth, downward motion of the mantle draws the oceanic crust back into the depths of the Earth—the process known as **subduction**. The geology of these regions is much more complex than that of the mid-ocean ridges. Figure 5.4 illustrates the kind of processes which take place when oceanic crust is subducted beneath the edge of a continent. This is happening, for example, along the western edge of the American continent and leads to many of the important geological features of this region: the formation of mountain chains such as the Andes, together with many active volcanoes and frequent earthquakes. This originally bare ocean crust has, by this time, acquired considerable sedimentary deposits. On sinking into the Earth, the basalt and sediment are heated, and some partial re-melting occurs. The rising magma may give rise to volcanoes, or solidify peacefully before reaching the surface. A common rock type of these regions is **andesite**, which is formed in this way. However, the rising magma from the melting ocean crust may well interact with existing continental crust on its way, causing remelting of this, together with mixing of different components. It is likley that

granite, a typical igneous rock of the upper continental crust, may often be formed by this more complex reworking process.

This rather oversimplified picture shows how the sea-bed crust is formed directly in the first stage of melting of the mantle, whereas the continental crust is much more complex, and often results from several stages of melting and solidification. These features are apparent in the chemistry of the different types of crust, that of the continents showing a more 'evolved' character, with more marked differences from the mantle. Some of the chemical changes will be described in the next section, but it is easy to appreciate that the processes of melting and crystallization of magma are rather similar to the recrystallization procedures used in the chemical laboratory for separating and purifying different compounds. The tectonic cycles are not only important in determining the gross geological features of the Earth's surface, but they also act as a gigantic chemical refinery, concentrating some elements in the newly-formed crust, and leaving others behind in the mantle.

Although the tectonic processes forming the crust are fairly well understood for the present-day Earth, there is still a considerable divergence of opinion about how they operated in the distant past. The oldest crustal rocks known are dated at 3.8 billion years, compared with the estimated age of 4.6 billion years for the Earth. (Chapter 6 explains how measurements of isotopic abundance are used to obtain these ages.) Most of the Earth's crust is much younger. This suggests that the continental crust, at least, has been accumulating rather gradually over the course of the Earth's history. Since the heat production shown in Fig. 5.3 has declined considerably with time, the mantle convection was probably more vigorous in the early stages. Thus, the rate of accumulation and recycling of the crust may well have been more rapid in the past, and have slowed down subsequently. There is geological evidence that major tectonic activity has not occurred continuously since the formation of the Earth, but rather has been concentrated in a number of cycles: the latest of these apparently started a few hundred million years ago and is continuing today. It has been suggested that these cycles may be related to changes in the pattern of convection currents in the mantle, which have occurred from time to time as the heat source has declined.

The chemistry of silicate minerals

All major rocks in the mantle or the crust are dominated by silicate minerals. The chemical changes resulting from the tectonic processes described in the previous section are, therefore, controlled to a large extent by the chemistry of silicates.

Except at extremely high pressures, all silicates are built from units, where a silicon atom is surrounded tetrahedrally by four oxygens (see Fig. 5.5). The isolated tetrahedron may be formulated as $(SiO_4)^{4-}$, and is found in **orthosilicate** minerals such as olivine, $(Mg,Fe)_2SiO_4$, and zircon, $ZrSiO_4$. In the structures of these minerals, as of the ones described below, the silicate groups are arranged so as to leave appropriately sized spaces for the metal cations. However, the tetrahedral unit has a strong tendency to polymerize, and this leads to a chemistry of considerable variety and complexity. The basic polymerization step may be written.

$$2(SiO_4)^{2-} \rightarrow (O_3Si{-}O{-}SiO_3)^{6-} + O^{2-} \qquad (5.1)$$

and gives a dimeric unit, consisting of two SiO_4 tetrahedra sharing a common corner oxygen. Minerals with this dimeric anion are known (for example, thortveitite $Sc_2Si_2O_7$), but they are rare, and the polymerization usually proceeds further, to form rings, chains, and more complex networks: these all have in common, however, the corner shared tetrahedra just described. Some of these structures are illustrated in Table 5.2, and show how progressive polymerization can lead to double chains, sheets, and finally three-dimensional frameworks in which all tetrahedra share corners with their neighbours, as in quartz SiO_2.

Table 5.2 also illustrates some of the important chemical trends that accompany the polymerization of the silicate group. Most obvious is the fact that the proportion of silicon, relative to oxygen and metal cations, increases as the polymerization proceeds. More subtle changes involve the nature of the other ions that accompany these structures. The least polymerized structures are commonly found in association with relatively small metal cations which form strong ionic bonds. With increasing polymerization the ionic charge on the silicate network decreases, and it becomes more favourable to incorporate ions which are larger (Ca^{2+} rather than Mg^{2+}) and of lower charge (Na^+ and K^+). Another important feature is that Al may replace some of the Si *within* the tetrahedral network. The different charge involved (formally Al^{3+} as against Si^{4+}) is then compensated by other ions, generally at larger sites in the structure. For example, Al may replace some Si in the SiO_2 framework, with the incorporation of Na, K, or Ca to form **feldspars**, the commonest mineral type in the crust.

When a mixture of molten silicates cools, the first minerals to crystallize are orthosilicates like olivine. These are succeeded by chain silicates (pyroxenes such as $MgSiO_3$) and thence by minerals with a progessively higher degree of polymerization. This sequence of

Table 5.2
Classification of major silicate minerals

Type	*Structure*	*Composition of tetrahedral groups*	*Si:O ratio*	*Mineral*	*Typical formula*
Separate tetrahedron	Plan view used below	$[SiO_4]^{4-}$	1:4	Olivine	$(Mg,Fe)_2SiO_4$
Chain		$[SiO_3]^{2-}$	1:3	Pyroxene	$MgSiO_3$
Double chain		$[Si_4O_{11}]^{6-}$	1:2.75	Amphiboles	$Ca_2(Mg,Fe)_5[Si_8O_{22}](OH,F)_2$
Sheet		$[Si_2O_5]^{2-}$	1:2.5	Clay minerals	
				—Kaolinite	$Al_4[Si_4O_{10}](OH)_8$
		$[AlSi_3O_{10}]^{5-}$		Mica	$K(Mg,Al,Fe)_{2-3}[AlSi_3O_{10}](OH)_8$
3-D framework	(SiO_4) tetrahedron sharing all vertices	SiO_2	1:2	Quartz	SiO_2
		$[AlSi_3O_8]^-$		Feldspars	
				—alkali	$(Na,K)[AlSi_3O_8]$
				—plagioclase	$(Na,Ca)[(Al,Si)_4O_8]$

crystallization means that a cooling magma will become progressively enriched in silicon, and in elements such as calcium and the alkali metals which have lower affinities for the simpler silicates. On the other hand, magnesium and iron (in the ferrous Fe^{2+} state) will accompany the phase which solidifies early on. Conversely, when a mixture of silicates is heated enough for partial melting to occur, the liquid formed will be rich in silicon and alkali elements, but relatively poor in magnesium and iron. With this in mind, one can understand the major trends in mineral types and chemical composition that accompany the formation of the Earth's crust. Figure 5.5 shows this trend, displaying both the main mineral constituents and elemental compositions, of rocks derived by progressive degrees of fractionation from the mantle: **basalt**, making up the

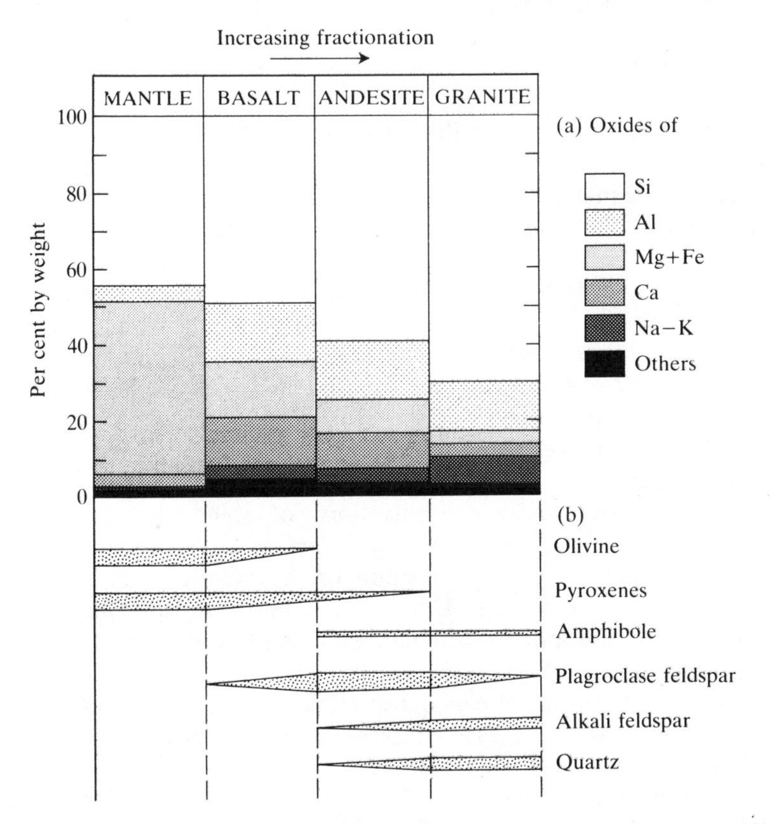

Fig. 5.5. Trends in the constitution of rocks derived by successively greater fractionation of the mantle. The upper diagram shows the break-down of the chemical composition in terms of oxides of the major elements; that below indicates qualitatively the proportions of different mineral types present in each rock.

ocean crust formed directly from the mantle; **andesite** and **granite**, formed in the continental crust by successive remelting as described in the previous section. Whereas the mantle is composed mostly of the 'simpler' olivines and pyroxenes, the highly polymerized framework silicates, feldspars, and quartz, come to predominate in the evolved continental rocks. These trends are reflected in the patterns of elemental abundance (expressed in Fig. 5.5, as in most geological data, as the weight percentage of simple oxides: this is a 'book-keeping' device, however, and does not imply that many of the elements are present as simple binary oxides!). The mantle is rich in magnesium and ferrous iron. With increasing fractionation, the proportions of silicon and the alkalis, sodium and potassium, increase.

In geochemical terminology, rocks with relatively low Si content are referred to as **basic**, or **mafic**, the latter term implying high Mg + Fe. On the other hand, crustal rocks with high Si (and Al) are called **acidic** or **sialic**. Thus, the major chemical consequence of tectonic processes is the formation of sialic crust from the mafic mantle. The fate of less common elements in this process is also important as it determines their availability in minerals at the Earth's surface. We have seen previously (p. 134 and Fig. 5.2) that elements with similar size and chemical behaviour may replace each other easily in minerals. For example, Cr and Ni can exchange for Mg in this way, and so tend to be retained by the mafic rocks in the mantle, and are less common in the crust. The fractionation of lithophilic elements between the mantle and crust shows a very strong correlation with ionic size and charge, displayed in Fig. 5.6. The contours in this plot show the crustal abundances relative to the total in the crust and mantle. The most strongly 'upwardly mobile' elements are those least similar to magnesium, either in size or charge. The high relative crustal concentrations of some elements in Fig. 5.6 – the alkalis, Be, Ba, Nb, Mo, Th, and U particularly—seems to indicate that quite a large proportion of the mantle must have been depleted of these elements. Detailed estimates suggest indeed that between a third and a half of the entire mantle has been fractionated in this way since the formation of the Earth.

Similar considerations of element distribution apply to the crust itself. Rare elements which can easily replace common ones in minerals are called **compatible** elements. Because they occur widely in low concentrations in many minerals, they are rather evenly spread over the Earth's surface. An example is gallium, which is found in most minerals containing aluminium, but almost never occurs in a concentrated form. On the other hand, many rare elements are **incompatible**: their ionic size or charge makes them difficult to substitute for common elements, so that their solubility in major minerals is small. Incompatible elements

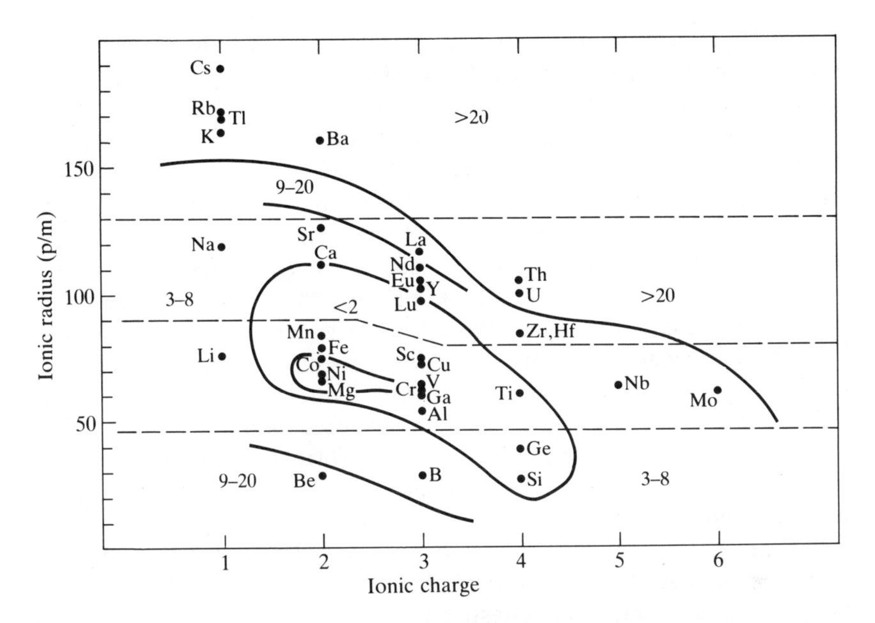

Fig. 5.6. Showing the enrichment of lithophilic elements in the continental crust. The values marked by the contours are (crust)/(crust + mantle) composition ratios, and are plotted against the normal ionic charge and radius of each element. It can be seen that crustal enrichment is controlled by the extent to which the radius or charge differs from that of the commonest mantle cation, Mg^{2+}. (From Taylor and McLennan 1985.)

pass easily into the liquid phase when magma is formed, and tend to remain in the silicate melt as it progressively crystallizes. They are, therefore, likely to be concentrated in the rocks which solidify last, where they may form minerals of their own. The list of incompatible lithophiles includes ones with small size (Li, Be, B) large ones (Cs, La and the lanthanides, Th, U), and ones which are commonly present in an unusually high oxidation state (Nb, Ta, W). A common source of some of these elements is the rock type known as **pegmatite**, a course-grained formation representing the last stage of solidification of sialic rocks such as granite. As will be described later, however, weathering processes may further modify and separate out the various minerals.

The formation of ores: hydrothermal processes

The ores of some elements may be formed directly during the crystallization of magmas as described in the previous section. Some oxides, such as those of iron and chromium, may solidify quite early in the crystallization sequence, and form **layered deposits** in basic rocks, often of quite ancient origin. Strongly siderophilic or chalcophilic elements are more likely to be present in the form of alloys or sulphides, which have low solubilities in the silicate melt. In some cases, these may also solidify at an early stage during the cooling of the magma. For example, platinum metals and sulphide ores are often found in layered deposits. The majority of ores are probably not laid down by direct crystallization from magma in this way, however, but are formed instead by **hydrothermal processes**: that is, by the action of water at high pressures and temperatures. Many minerals, normally thought of as highly insoluble in water, can dissolve to form hydrothermal solutions, and are thus transported through the crust and finally precipitated in concentrated ores. Such processes are important in the formation not only of many sulphide ores and native metal deposits, but also of oxide minerals of some elements such as Sn and W.

Many aspects of the origin of hydrothermal deposits are not understood, in spite of a great deal of research in recent years. There are three particularly important questions.

1. Where does the water originate?
2. In what chemical form do different elements dissolve in the hydrothermal fluid?
3. What is the mechanism of precipitation of the ore?

All magma contains some water, which can be an important ingredient in igneous processes. It lowers the melting temperature of rocks, and decreases the viscosity of the magma so that it rises and flows more easily. Water will become progressively concentrated in the liquid phase as the magma starts to solidify, and violent volcanic eruptions sometimes occur as it boils when the magma nears the surface and the pressure is reduced. Some of this water may come from the decomposition of hydrated minerals. However, isotopic studies suggest that a high proportion of hydrothermal water in fact originates from the surface, as rain or ocean water which percolates deep into the crust and becomes superheated under pressure.

A detailed understanding of the chemical species present in hydrothermal solutions is difficult because of the high temperatures and pressures involved, which are well outside the range of most laboratory studies. In the case of lithophilic elements, the solubility of their oxides may be quite high at elevated temperatures. This is well established for SiO_2, which dissolves as $Si(OH)_4$, leading to the common occurrence of quartz in hydrothermal deposits. Other elements such as Sn and W may be present as similar hydroxide complexes, or oxyanions like WO_4^{2-}. Elements with low lithophilic character, however, are unlikely to be sufficiently soluble in this form. Hydrothermal fluids contain a range of anions, such as chloride and sulphide, with which chalcophiles and even siderophiles are known to form strong complexes. Thus, gold, which cannot dissolve as a simple cation, may form the complex ions $AuCl_2^-$ and $Au(SH)_2^-$. Some information about the composition of hydrothermal solutions can be obtained by studying fluid inclusions found in minerals such as quartz and thought to be remnants of the solution trapped during the solidification process. Chloride and sulphide ions are indeed found, together with others such as carbonate and fluoride; these latter probably increase the solubility of lithophilic metals by forming complexes with them.

The way in which the ores are actually deposited from solution is probably the most complex question of all. The geology and chemistry of the rocks surrounding the deposit sometimes provide clues, but very often these rocks have been modified by later geological changes, so that the original conditions have been obliterated. In a few cases the ore forming processes can be studied in action today. For example, **hydrothermal vents** occur on the sea floor, where streams of water with a temperature of 300°C or more emerge from the sea-bed. In appearance, these vents resemble smoking chimneys, as metal sulphides are precipitated where the hydrothermal solution comes into contact with sea water. The precipitation results both from cooling and from the lowering in concentration of the complexing anions. Massive deposits of highly concentrated sulphide ores may be formed from such hydrothermal vents. Later they may be covered with sedimentary deposits, which through geological changes eventually end up on dry land. The historically important copper deposits in Cyprus—from which the element derives its name—were probably formed in this way.

When the superheated solutions reach the Earth's surface the reduction in pressure causes the water to boil, resulting in hot springs or more violent volcanic eruptions. Many borate deposits appear to have been associated with hot springs. Sulphide ores such as cinnabar (HgS), as well as native sulphur, and deposits of fluorides and chlorides, can all be found near active and extinct volcanoes.

More frequently, ores probably result from the reaction of hydrothermal fluids with sedimentary and other rocks close to the Earth's surface. Sometimes this reaction apparently happens rapidly, resulting in a highly localized deposit at the interface between the original solution and the reactive rock. Carbonates are particularly prone to this type of process, which forms so-called **skarn deposits**. Reaction of acid water with carbonates raises the pH:

$$2H^+ + CaCO_3 \rightarrow Ca^{2+}(aq.) + H_2O + CO_2. \quad (5.2)$$

This, in turn, increases the dissociation of the weakly acidic H_2S, which can lead to the precipitation of metal sulphides:

$$M^{2+}(aq.) + H_2S \rightarrow MS + 2H^+. \quad (5.3)$$

Other elements can be deposited by direct replacement of ions in the original mineral, for example:

$$CaCO_3 + WO_4^{2-} \rightarrow CaWO_4 + CO_3^{2-} \quad (5.4)$$

The reaction with surrounding rock may occur more slowly, and allow the hydrothermal solution to percolate over a wide area, with different metals precipitating at different distances from its origin. This leads to the frequent **zoning** of ores. For example, granite formed by igneous processes in south-west England is surrounded by metal-containing deposits occurring in four zones, containing the minerals shown below:

(1) SnO_2;

(2) $FeAsS$, $FeCuS_2$, $(Fe,Mn)WO_4$;

(3) PbS, ZnS, Ag_2S;

(4) Fe_2O_3, $FeCO_3$, $MnCO_3$.

The first of these zones is closest to the granite and, hence, presumably to the origin of the hydrothermal solution. Similar zoning is found in many ore deposits of hydrothermal origin, although the precise ordering of the minerals within the zones is not always the same. The sequences must depend both on the original concentrations of the different elements in solution, and on the chemical processes which result in precipitation. In some cases, a correlation can be found between the solubilities of the different sulphides and their position in the sequence.

The results of hydrothermal processes can often be seen on a small scale, in rocks on the coastline or in other exposed places. Many rocks, especially igneous ones such as granite, often contain veins or inclusions of crystalline quartz or calcite, deposited by hydrothermal solutions which have percolated through cracks and cavities in the host rock. In some cases, crystals of ore minerals such as pyrites, FeS_2, chalcopyrite, $FeCuS_2$, or galena, PbS, can be found along with the commoner minerals.

Since hydrothermal solutions are often of igneous origin, it is not surprising that the resulting ore deposits are found in places which either show considerable tectonic activity at present (as with the Andes region of South America) or where there is evidence for such activity in the past. However, both the geological origin of ores and their geographical location still present many mysteries. Many deposits are concentrated in particular regions, for example, in certain areas of the African continent, known as **metallogenic provinces**. It seems as though certain times in the past, when these parts of the crust were tectonically active, were especially efficient at producing metal-containing ores. Different elements show different distributions in time as well as in space: for example, deposits of nickel and chromium seem to have been formed rather early in the development of the crust, whereas ores of many other elements, such as copper, tin and molbydenum are more recent. To some extent, these differences must reflect the chemical behaviour of the elements through the rather complex sequence of tectonic cycles which have made the continental crust. It has also been suggested that they may indicate an uneven distribution of some elements in the mantle, although such a hypothesis is not easy to test.

Weathering and the formation of sedimentary deposits

Weathering is the action of water and air on surface rocks. It causes both physical and chemical modification of igneous rocks, and continues the process of redistribution and fractionation of the elements that is the main theme of this chapter.

The most important overall chemical consequence of weathering is the formation of hydrated silicates, especially the **clay minerals**, which have structures based on two-dimensional silicate layers (see Fig. 5.5). This type of structure is stabilized by the presence of water, which (as OH^-) forms hydrogen bonds between the layers. A typical weathering reaction is the action of CO_2—saturated water on potassium feldspar, to give kaolinite and hydrated silica:

$$4KAlSi_3O_8 + 4CO_2 + 22H_2O \rightarrow 4K^+ + 4HCO_3^- + Al_4Si_4O_{10}(OH)_8 + 8H_4SiO_4 \quad (5.5)$$

Even this reaction is far from simple, and given the wide range of minerals that exist in most igneous rocks, it is clear that the chemistry of weathering processes is extremely complex. There are some general principles, however, which are important in understanding the redistribution of elements which results from weathering. In the first place, minerals differ very widely in their resistance to weathering processes. With silicates, this is correlated with the Si/O ratio, and it is found that the less polymerized structures, such as olivine and pyroxenes, are the most susceptible to attack. Quartz is extremely resistant, and among three-dimensional framework silicates the calcium- and sodium-rich plagioclase feldspars weather more rapidly than potassium feldspar. (This latter difference is probably related to the lower hydration energy of the larger K^+ ion.) This order of stability is clearly important in controlling the release of different elements into solution. Thus, magnesium, calcium, and sodium are more easily released into solution by the decomposition of rocks than is potassium.

The solubility of different elements in water is clearly another important factor in the chemistry of weathering. For many elements, solubility is a function of the pH and the presence of anions such as carbonate, but a useful general guide to aqueous solubility of an element at neutral pH is provided by its **ionic potential**, defined as the formal charge divided by the ionic radius. Figure 5.7 shows a plot of charge against ionic radius for several elements with common positive oxidation states. The heavy lines through the origin separate three groups. Ions with Z/r less than 0.03/pm are generally soluble at a pH close to 7, forming simple hydrated cations. Ones with Z/r greater than 0.12 form soluble oxyanions: some of the elements in this category in Fig. 5.7 are non-metals, for which concepts of 'ionic charge' and 'ionic radius' have only a formal significance (see Chapter 2, p. 45). The least soluble elements at neutral pH are those with intermediate values of ionic potential, occurring between the two lines on the diagram. These elements form oxy and hydroxy species, which are often highly polymerized and insoluble, although they may dissolve more readily under acidic or basic conditions: for example, silicon is more soluble at high pH, and amphoteric elements, such as aluminium, can dissolve in both acidic and alkaline solutions. It can be seen that the position of some elements also depends on their oxidation state. Iron and manganese are relatively soluble in the 2+ state, but the oxidized forms Fe^{3+} and Mn^{4+} are highly insoluble.

Using these general guidelines, one can understand the kinds of chemical changes which take place in rocks during weathering. The changes are shown most clearly under extreme conditions, for example, in tropical climates where water is abundant and high temperatures speed up the chemical reactions. Alkali metals and those of group 2 (except Be) will be washed out of rocks, as will many of the non-metals which form anions (this includes, of course, the halogens, particularly chlorine and bromine which occur as very soluble simple anions). Iron and manganese present as 2+ ions (which is the case in many igneous rocks) will also go into solution, although subsequent oxidation by air will result in their precipitation. Some elements, on the other hand, will remain as solids. Minerals which are unaltered by weathering may include SiO_2 (quartz), oxides of Ti and Sn, and the lanthanide mineral monazite, $LnPO_3$. These are classed as **resistates**. The ultimate dissolution of clay minerals produces **hydrolysates**, which include the very insoluble element aluminium as the hydrated oxide such AlO(OH).

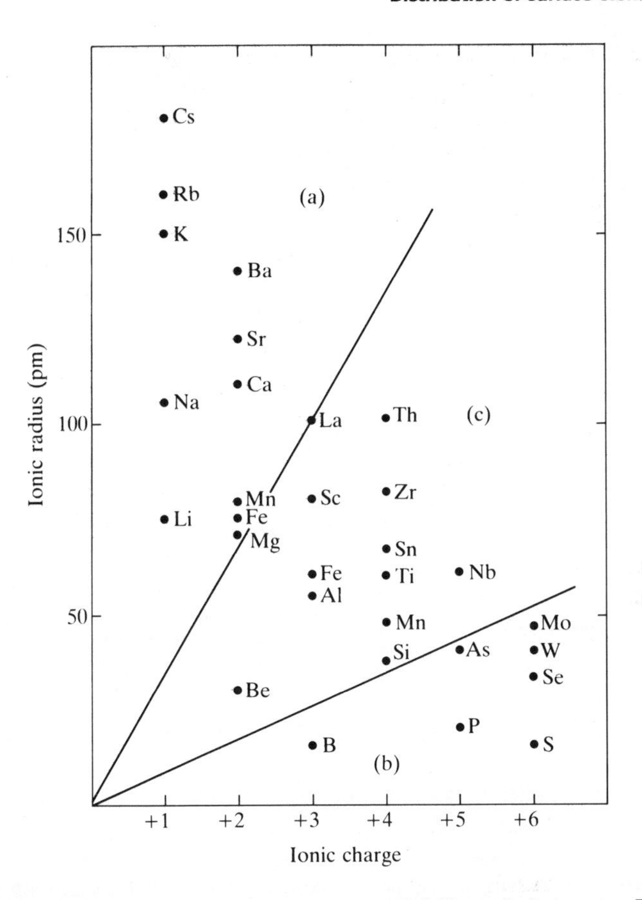

Fig. 5.7. Ionic potential and the aqueous solubility of elements. The solid lines divide (a) elements with $Z/r < 0.03$/pm, which form soluble hydrated cations such as Ca^{2+}; (b) ones with $Z/r > 0.12$/pm, soluble as oxyanions such as SO_4^{2-}; and (c) those of intermediate Z/r, which form oxides or hydroxides insoluble around neutral pH.

The chemical processes resulting from extreme weathering are summarized in Fig. 5.8. It can be seen that there may be considerable simplification of the rock chemistry, with the loss of soluble elements leaving others separated as simple compounds. Many important mineral deposits have been formed in this way, including **bauxite**, made up of hydrated forms of aluminium oxide, and iron ores such as **haematite** (Fe_2O_3), which can result from the oxidation of Fe^{2+} released in weathering. Other elements, such as titanium, zirconium, and the lanthanides, are often found in conjunction with silica, for example, occurring as beach sand.

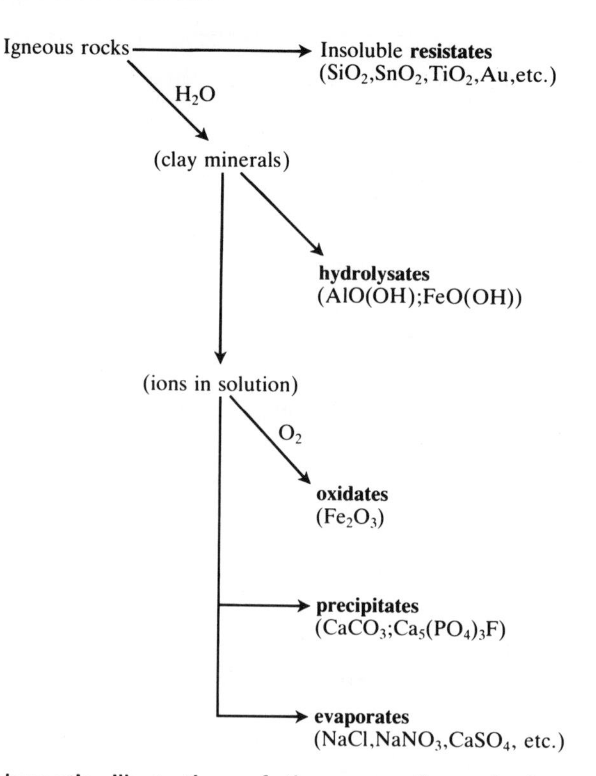

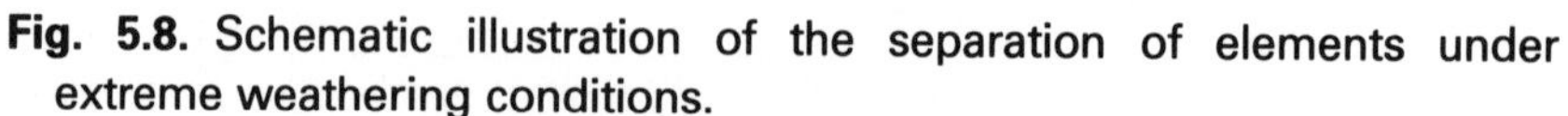

Fig. 5.8. Schematic illustration of the separation of elements under extreme weathering conditions.

The fate of the soluble elements removed from rocks during weathering depends on their relative solubility, especially in conjunction with anions also present, such as carbonate and sulphate. Many of these elements pass into the ocean, and will be considered in more detail in the following section. However, we can note that some elements, such as calcium, precipitate fairly easily, especially as the carbonates which form such widespread sedimentary rocks. Other more soluble elements such as sodium may remain in solution for a very long time, although they eventually form **evaporite** deposits in regions where enclosed seas or lakes are dried up by evaporation.

Some sedimentary processes have changed quite markedly over the course of the Earth's history. They illustrate how the fates of different elements may be bound together in quite an interesting way. For example, the formation of sulphate minerals, common for calcium, strontium, and barium, cannot have occurred until the Earth acquired an oxidizing atmosphere through the development of photosynthesis, as before that time sulphur was only present in the reduced sulphide form.

A more dramatic illustration is the existence of **banded ironstone formations**, so called because iron oxide layers alternate between other sedimentary minerals. These formations were laid down between 2 and 3 billion years ago, at a time when the oxygen content of the atmosphere was starting to increase. Before this time, iron (as Fe^{2+}) was a relatively soluble element, and must have been abundant in the oceans, whereas under present atmospheric conditions oxidation to the insoluble Fe^{3+} occurs rapidly. The oxidation of Fe^{2+}, which led to the banded ironstones, formed a very important 'sink' for free oxygen, and must have kept its atmospheric concentration at a low level for perhaps as long as a billion years. Uranium is another element for which the solubility is strongly dependent on the availability of oxygen, and this factor seems to have been important in the formation of the remarkable Oklo deposit, discussed in Chapter 6.

Weathering processes produce important physical changes in rocks, as well as chemical ones. The mechanical breakdown of rocks by the continued action of water and ice is essential in allowing the chemical reactions to proceed to completion. Once formed, the chemically resistant minerals may be mechanically sorted, according to their different density and particle size, by running water in streams. Ores which have been separated in this way are called **placer deposits**. These can include not only oxide minerals such as SnO_2, but also highly resistant native metals, such as gold and the platinum metals. Most earlier sources of gold, including those worked in the 'gold rushes' in the nineteenth century, were of this kind, although these surface deposits have now been mostly exhausted.

The elements in the ocean

The oceans cover two-thirds of the Earth's surface and contain—in the form of water—a high proportion of the available hydrogen on Earth. They are also major reservoirs of the halogens Cl, Br, and I. For a number of soluble elements, processes involving precipitation from the ocean, or evaporation of sea water, have led to the formation of concentrated deposits. Thus, the ocean forms an important part of global cycles by which some elements are redistributed, and is the medium from which sedimentary deposits such as $CaCO_3$ have mostly been laid down. It is also important as the environment in which life on Earth almost certainly originated and, indeed, biological processes still exert a controlling factor on the distribution of many elements in the ocean.

Studies of the element distribution in the ocean have undergone major developments in recent years. Advances in analytical techniques, together with sampling methods designed to avoid contamination, have led to a revision in the estimated concentration of a number of elements. There has also been an increasing recognition that many elements show important changes in concentration with depth in the ocean. Another problem is the presence of large amounts of suspended matter, for example, clay minerals washed out by rivers, which may add considerably to the overall abundance of elements such as aluminium and iron, especially in coastal regions.

Figure 5.9 shows the estimated average concentration of dissolved elements in sea water, plotted in two ways. In Fig. 5.9(a) the concentration is given on a logarithmic scale extending down to below 10^{-12} mol/kg. In Fig. 5.9(b), the average sea water abundance is compared with that of elements in the crust. As in the previous plots of this kind (see Figs 4.6 and 5.1), the abundance ratio of elements is shown against their group in the periodic table. As the crust is undoubtedly the major source for most elements, Fig. 5.9(b) illustrates the effects of the fractionation processes which distribute elements between solid rock and aqueous phases. The predominant type of chemical species present in sea water is also shown for each element. In most cases, this information comes not from direct measurements on sea water itself, but is deduced from laboratory measured data on the equilibrium constants involving the different anions and cations known to be present. It can be seen that only a minority of elements—comprising halogens F, Cl and Br, and a number of metal ions with a charge no greater than 2+—are present as simple hydrated anions or cations. Most elements with higher oxidation states form hydroxy- or oxy-complexes, ranging from neutral ones such as $Fe(OH)_3$, through anionic species like $SnO(OH)_3^-$, to oxyanions such as SO_4^{2-} and MoO_4^{2-}. Many other types of complexes can be formed, the predominant ligands being carbonate or chloride ions.

The trends shown in Fig. 5.9 obviously reflect to a large extent the patterns of solubility discussed in the previous section, and summarized there in Fig. 5.7. The elements present in the highest relative proportion in the ocean are those which are soluble either as cations (low charge/radius ratio), or as oxyanions (high formal charge), together with the very soluble halide ions. On the other hand, metallic elements occurring in the oxidation states 3+ and 4+ have particularly insoluble oxides and hydroxides, and have very low relative abundances. This is especially striking for aluminium, titanium, and iron, which are abundant elements in the crust, but have extremely low dissolved concentrations in the oceans. A few metals may seem to have suprisingly high relative abundances, but again, this can be understood from their chemical behaviour. For example, uranium and molybdenum form the very soluble oxy- species UO_2^{2+} and MoO_4^{2-}, respectively. In the case of gold, its

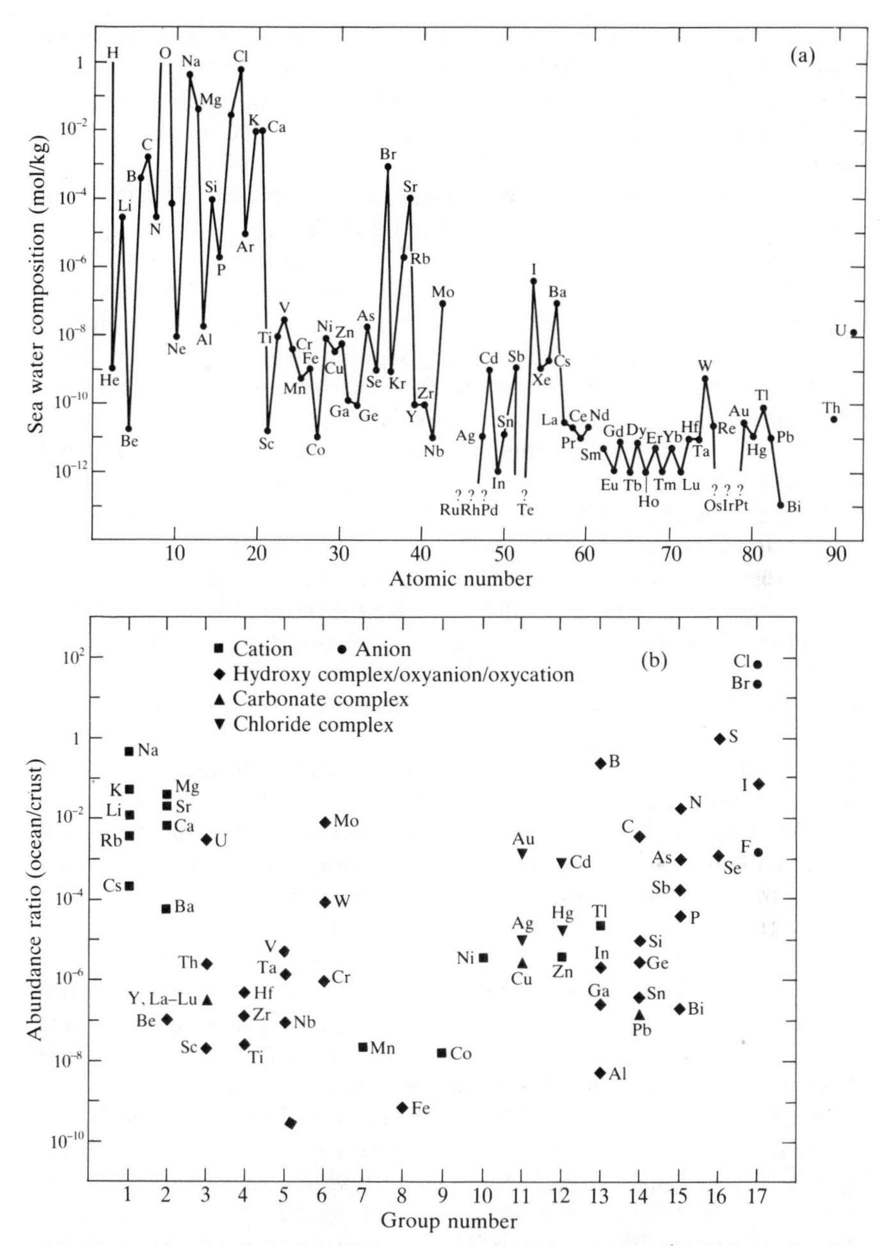

Fig. 5.9. Elemental abundances in the oceans. (a) Concentration of dissolved species in sea water of average salinity (35 parts per thousand of halide ions). (b) Comparison of abundances between the ocean and the continental crust. The diagram also shows the major species present. (See Appendix A for data on abundances; species present from Henderson 1982.)

relatively high position in Fig. 5.9 must be related to the stability of the soluble chloride complex $AuCl_2^-$.

In spite of the generally satisfactory correlation of elemental abundances with their chemical behaviour, many detailed aspects of the chemistry of sea water, and especially of the mechanisms which control the abundances, are not understood. Most metallic elements are probably derived from the weathering of continental rocks, although sea-bed processes involving the action of water on the ocean crust or through hydrothermal sources deep in the crust, are also important. For example, it is established that a major source of Mn is from hydrothermal vents along the mid-ocean ridges, and the same may be true for a number of other elements. Some non-metallic elements such as Cl, S, and Br, however, are probably derived mostly from a different source. The concentration of these elements in the crust is thought to be insufficient to account for their abundance in the ocean, and volcanic sources, particularly in the early stages of formation of the atmosphere by outgassing of minerals, may be the most important origin.

Although the source of elements is important, it is quite wrong to think of the ocean as a 'sink' in which they have been slowly accumulating. Estimates of the rate of input of many elements show that they would quickly reach concentrations much greater than those observed. There must, therefore, be processes by which elements are removed from the ocean, the observed abundances being determined by a balance between the rates of addition and removal. This balance leads to the concept of a **residence time**, representing the average time that atoms of a given element spend in the ocean before being removed. If the total amount A of an element in the ocean, and its rate of addition (or removal) $\dot{A}$ are both known, then the residence time, t, can be defined as:

$$t = A/\dot{A}. \tag{5.6}$$

Some estimates of residence times are shown in Fig. 5.10, plotted against the ratio of the ocean to crustal abundances from Fig. 5.9(b). Since there are large uncertainties in the rates of input and output, these residence times must be regarded with caution, and represent order-of-magnitude estimates only. Nevertheless, there is quite a striking correlation shown in the diagram. Elements with high relative abundance (Na, B, Cl, Br, etc.) have very long residence times of around 10^8 years, whereas those with very low relative concentration, such as Al and Fe, have much shorter values of 100 years or so. This strongly emphasizes the point that the abundance is controlled at least as much by the rate of removal of elements from sea water as by the rate of addition.

In spite of the obvious importance of processes which remove elements from the ocean, the nature of these is generally poorly under-

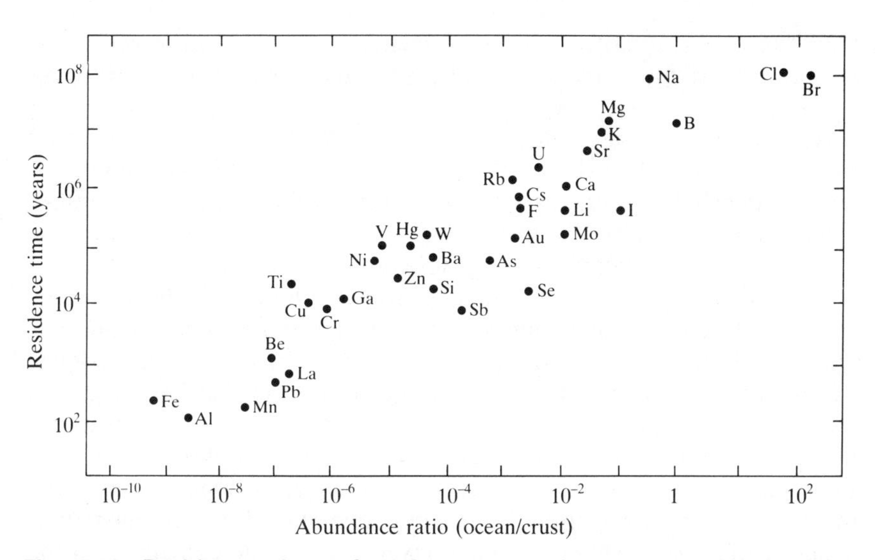

Fig. 5.10. Residence times for elements in the oceans, showing the correlation with their ocean/crust abundance ratios from Fig. 5.9.

stood. For very soluble halide ions, the residence times are of the same order as those of geological processes which move the continents and alter the crust. They are removed predominantly by evaporation of water from seas which become enclosed and cut off from the rest of the ocean. **Evaporite** deposits formed in this way are the main source of Cl and Br. Although originally formed at the surface, they are often buried by other sedimentary deposits, and are found today underground. In a few other cases, elements are removed by well-established precipitation reactions. For example, surface waters are slightly supersaturated in calcium carbonate. Under inorganic conditions, and especially in the presence of magnesium, precipitation would be very slow, but calcium carbonate is used by many marine organisms to make hard skeletons, and it is these biological processes which lead to the removal of calcium. The dead organisms finally sink to the sea-bed, giving the limestone deposits which form an important fraction of sedimentary rocks. Many other, less abundant elements may also be removed by marine life. This often has the consequence that surface water, where such life is concentrated, is relatively depleted in a given element. Elements in this category are said to have **nutrient-type distributions**, and include some that are known to be essential to life, such as C, N, P, Ca, Si (which is used to make SiO_2 skeletons), Cu, Zn, Se, and I. However, a number of other elements seem to be taken up by living organisms in surface waters, including Ag and Cd. Such biological take-up may not always provide the ultimate mechanism for removal of the elements concerned, however, as it appears that some nutrient-type elements are released again by dying organisms as they sink.

Another interesting removal mechanism leads to the formation of **ferro-manganese deposits** on the ocean floor. Often known as 'manganese nodules', these contain hydrated oxides of Fe, Mn, and many other elements in economically interesting concentrations. The controlling factor in the precipitation of manganese seems to be the rate of oxidation from soluble Mn^{2+}—the predominant form in solution—to the much less soluble 3+ and 4+ oxidation states. Although this reaction is thermodynamically favourable in oxygenated water, it occurs extremely slowly except in the presence of a catalyst, and it appears that hydrated iron oxides are efficient in this respect. The incorporation of other elements into these deposits, however, is less well understood. Indeed, it is true to say that for many of the elements with short residence times, the mechanism of removal from the ocean is uncertain. The concentrations of many of these elements is much lower than can be explained on the basis of precipitation of the least soluble known compound. It is likely that adsorption of ions from solution, onto the surface of clays and other mineral particles, is sometimes the major factor.

The elements of life

So far in this chapter we have looked at the fractionation of the elements into the Earth's crust, into the different minerals which make this up, and into the ocean. One further stage of fractionation is worth considering, as it is important to us in a more intimate way: this is the selection of elements by living organisms. Since life very probably originated in the oceans, and was certainly confined there until quite recently on a geological time scale, it seems reasonable to compare the elemental abundances of living matter with those in the ocean. This is done in Fig. 5.11, by plotting the composition of the human body (expressed, as is normal for biological samples, in terms of the **dry mass** after removal of water) against that of dissolved species in sea water. There is an enormous scatter on this diagram; nevertheless, the correlation is significant. Most elements with a sea water concentration greater than 10^{-6} mol/kg are relatively abundant in the body and are essential to life. Rarer elements in the ocean are present in life in trace amounts only and, in many cases below the level of detection and so not plotted. Relative concentration or depletion of elements may also be judged from the diagram. Those nearer the top left-hand corner are concentrated in the body. They include the major biological elements carbon, nitrogen, and

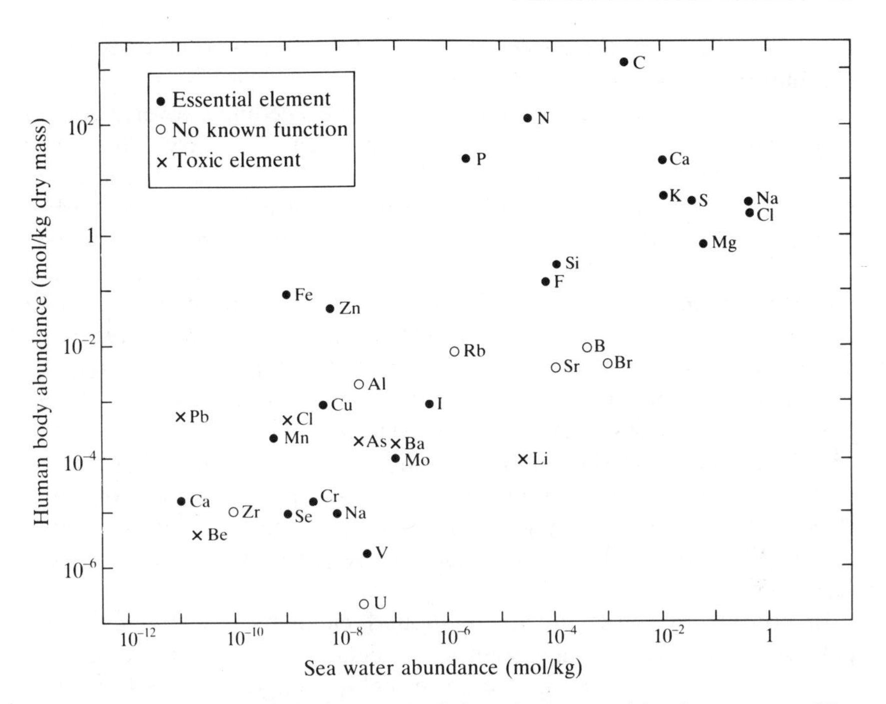

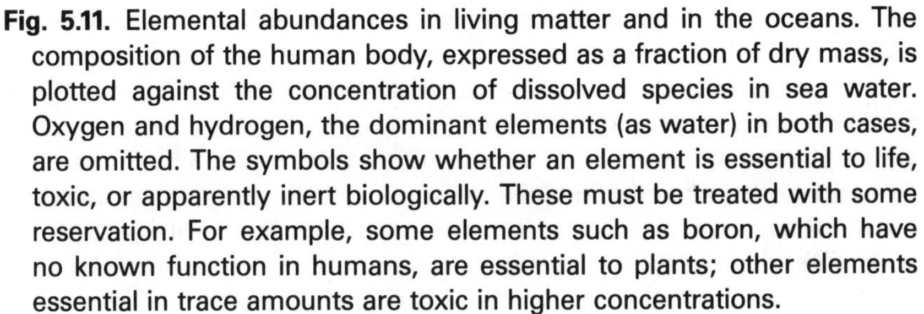

Fig. 5.11. Elemental abundances in living matter and in the oceans. The composition of the human body, expressed as a fraction of dry mass, is plotted against the concentration of dissolved species in sea water. Oxygen and hydrogen, the dominant elements (as water) in both cases, are omitted. The symbols show whether an element is essential to life, toxic, or apparently inert biologically. These must be treated with some reservation. For example, some elements such as boron, which have no known function in humans, are essential to plants; other elements essential in trace amounts are toxic in higher concentrations.

phosphorus, as well as the minor, but exceedingly important ones, iron and zinc.

Figure 5.11 also distinguishes between those elements which are believed to be essential for human life, those which are toxic, and ones which apparently fulfil no known role. This distinction must be treated with some reservation. Some elements are known to be essential for other forms of life, for example, boron in plants, and strontium and barium in some marine organisms. Other elements, such as cobalt, selenium, and nickel, are essential to humans in trace amounts, but toxic in higher concentrations. There may even be some elements whose function is still unrecognized. For example, silicon and selenium are essential, but their function in animals is unknown; and there has recently been much dispute about the possible function—and toxicity—of aluminium.

The roles played by essential elements are exceedingly diverse and, in many cases, reflect well-known aspects of their chemistry. The major element is, of course, carbon, which forms a uniquely complex series of **organic compounds**, including the major 'molecules of life': carbohydrates (sugar, starch, etc.), lipids (fats), proteins (including enzymes), and nucleic acids (DNA, RNA). Some of these compounds have a structural function; others act as energy sources or stores, contain genetic information, or are catalysts which direct the chemical reactions of a living cell. In these molecules carbon forms covalent bonds, to other carbon atoms, and to hydrogen, oxygen, nitrogen, and sulphur. Phosphorus is equally essential, but its behaviour is slightly different, as it is present in the oxidized phosphate form, bonded to oxygen rather than to carbon directly. As such, phosphate plays an important role both as a structural component of nucleic acids, and in the reactions by which living cells utilize the energy sources required for their chemical reactions.

In contrast to these covalently bound elements, sodium, chlorine, potassium, and some magnesium and calcium, are present as charged ions—Na^+, Cl^-, and so on. The more highly charged cations Ca^{2+} and Mg^{2+} are often associated with negatively charged groups on organic molecules, such as proteins and nucleic acids. The ionic interactions may help to control the shape of these molecules or to activate their reactions in other ways. Other ions are 'free' in solution, forming an electrolyte which plays some essential, but quite subtle roles in life. One function is to control the **osmotic balance** between a cell and the surrounding medium, for example, body fluids or sea water. It is notable that the sodium content of living organisms is quite low, being less than that of potassium; in sea water, on the other hand, sodium is over 50 times more abundant than potassium. Within most cells the Na/K ratio is even lower than in the body as a whole. It seems that very early living cells probably developed the ability to exclude sodium in this way. If they did not do this then the osmotic pressure, having contributions from many organic solutes not present in the ocean, would be high enough to cause the cells to take in water, and so cause them to swell and burst. Being able to control their ionic content, many living cells have come to use this trick for other purposes. Concentration gradients of ions can give rise to electrical potential differences across cell membranes. Probably, all cells have such membrane potentials, but they are made particular use of in specialized cells such as nerves. The 'resting potential' of a nerve cell is associated with an abnormally low sodium concentration inside. 'Triggering' the cell at a synapse or nerve end causes the membrane to become locally permeable to sodium ions, which enter the cell and depolarize the membrane. By a mechanism which is not understood, this region of sodium permeability and depolarization passes along the nerve cell, thus transmitting the 'message'. The cell is then restored to its active condition by 'pumping' the sodium ions out again.

Another major use of some inorganic elements is in making hard skeletons. Calcium carbonate is the favourite mineral of many marine organisms for this purpose, and we have already mentioned how it contributes to the precipitation of calcium from the oceans, and to the formation of extensive sedimentary deposits. Silica (SiO_2) and more rarely strontium sulphate ($SrSO_4$) are other insoluble minerals used in the same way. On the other hand, animal bones and teeth are made from calcium phosphate in the form of hydroxyapatite, $Ca_5(PO_4)_3(OH)$.

Other elements play a more specialized function in life. Well known are the occurrence of magnesium in chlorophyll—the light-absorbing pigment essential for photosynthesis by green plants—and iron in haemoglobin, which transports oxygen in the blood stream. Iron and zinc, as well as other transition elements essential in trace amounts, are components of many enzymes. These protein molecules act as catalysts for biochemical reactions, and the role of transition metals here obviously has some parallel in the use of the same elements as catalysts in chemical industry and in the laboratory. It arises partly from the ability of the metals to co-ordinate molecules in ways which promote specific chemical reactions. Molybdenum probably acts in this way in the important process of **nitrogen fixation**, the conversion of the unreactive N_2 molecule from the atmosphere into a biologically useful form. In other cases, it is the existence of more than one stable oxidation state which is important. For example, cytochromes and ferredoxins are iron-containing enzymes which make use of the conversion between Fe^{2+} and Fe^{3+} forms. These enzymes form part of the energy utilizing reactions of cells, for example, in photosynthesis, and in the oxidation of food molecules by air-breathing organisms such as ourselves.

The availability of essential elements is often a crucial factor in limiting the extent to which a given environment can support life. This is true in the oceans, where, as we have seen, several elements are strongly depleted in surface layers as a result of their uptake by marine life. The enormous fertility of certain areas, such as the Pacific coast of South America, is attributed to the abundance of inorganic nutrients brought to the surface by upwelling currents. On land, most necessary minerals are derived from the soil. The presence of adequate phosphorus, potassium, and other elements required by plants is well recognized by farmers and gardeners. Element deficiencies, for example, of iron, zinc, iodine, and selenium, may also play a role in human disease.

Several elements are highly toxic, not only to humans, but to most forms of life. The 'heavy metals', cadmium, mercury, and lead, are especially notorious in this respect. It is notable that these elements are all chalcophiles, with a strong chemical affinity for sulphur. This is, indeed, thought to be the basis of their toxicity, as they can combine with sulphur in enzymes, and either replace an essential element such as zinc or disrupt the function of the enzyme in some other way. These elements are present in low concentrations in the natural environment, but their concentration has increased markedly in recent years as a result of industrial processes. Most lead in the human body comes from sources such as automobile exhaust, drinking water contaminated by lead pipes, and paint. These are causes for concern, especially as the heavy metals accumulate in the body over long periods; thus, their toxicity may have subtle aspects not easily recognizable from studies covering a limited time-scale.

To end this section we shall look very briefly at the ways in which life has contributed to the distribution of elements on the Earth's surface. We have already mentioned the formation of sedimentary minerals such as calcium carbonate. The role of living organisms is probably only secondary here, speeding up a thermodynamically favourable process which otherwise would occur by purely inorganic means. More remarkable is the existence of elements in forms which are not thermodynamically stable ones, and so could not easily be produced without the free energy of living processes, derived ultimately from the Sun's radiation by photosynthesis. The result of photosynthesis is the conversion of water and carbon dioxide into organic (reduced) carbon and free oxygen. The unique composition of the Earth's atmosphere, and the occurrence of hydrocarbon deposits such as oil and coal, are thus the result of some billions of years activity by life on the Earth. The human use of hydrocarbon fuels is, of course, reversing this process; the depletion of these reserves, and the increase of atomspheric carbon dioxide which results from burning fossil fuels, are worrying. Especially serious may be the **greenhouse effect** resulting from the strong infra-red absorption by carbon dioxide, which therefore decreases the proportion of the Sun's radiation re-emitted from the Earth back into space. It is claimed that the increase in the atmospheric CO_2 content due to human activity may already be giving rise to irreversible—and largely deleterious—changes in the global climate.

Another element of interest here is sulphur. In some oxygen-poor environments such as ocean basins (the present Red Sea is an example), bacteria are able to obtain supplies of oxygen by the reduction of sulphate ions (SO_4^{2-}) to sulphides. This leads to the precipitation of some chalcophile elements as sulphides, and is probably a factor in the generation of some sulphide ores. In other circumstances, sulphide is converted by bacteria to elemental sulphur. Many of the world's reserves of this element (again in a thermodynamically less stable form) derive from such a source.

Summary

Following the initial segregation of the Earth into a metallic core, and an oxide mantle and crust, redistribution of elements has continued throughout geological history. Tectonic processes, fuelled by heat from the radioactive decay of a few elements, cause slow convection within the mantle. Partial melting of mantle rocks close to the surface forms the crust, a thin 'scum' which essentially floats on the denser mantle. The mantle is rich in magnesium silicate, and **compatible** elements, which easily enter this phase, tend to be retained there. These include chromium, nickel, and iron in the Fe^{2+} form. The crust is richer in silicon, and other elements, such as alkali metals and calcium, which enter the liquid phase when partial melting occurs. Highly **incompatible** elements are those which are not easily included in the mineral structures formed by common elements and may become concentrated in rare minerals. These include Li, Be, B, Cs, La, Th, U, Nb, Ta, and W.

The process of element fractionation is continued by the action of water. **Hydrothermal processes** take place at high tempertaures and pressures, and lead to the formation of many economically important oxide and sulphide ores, for example, of tin, copper, and molybdenum. **Weathering** and **sedimentary** processes take place at the surface, and produce a separation of elements according to the solubility of their compounds in water. Soluble elements such as calcium and sodium pass into the ocean, and eventually form sedimentary deposits by precipitation or evaporation. Other elements, such as aluminium, titanium, and iron, remain as highly insoluble oxides.

Nearly all known elements have been detected in dissolved form in the ocean. The enormous range of abundance, down to 10^{-12} molal concentration, largely reflects their range of solubility. However, the composition of the ocean results from a dynamic balance of processes of addition and removal; in many cases, both the predominant source of elements and the mechanisms by which they are finally removed from sea water are still uncertain.

Living organisms have their own, highly atypical element distribution. It is dominated by carbon and other elements present in organic compounds. However, a wide range of inorganic species are used by life for different purposes and are essential nutrients. The presence of life on our planet has itself influenced the distribution of elements in some unusual ways; for example, the high concentration of free oxygen in the atmosphere and the existence of fossil fuel deposits of organic carbon are due to this source. More recently, human activity is altering the distribution of some elements in ways which may have serious consequences for the future.

Further reading

Goldschmidt (1954) is a classic work on geochemistry, by a founder of the subject. It is still worth looking at, although now out of date in many ways. The best modern account is that of Mason and Moore (1982) (or Mason alone in previous editions). Fergusson (1982) gives a wide-ranging discussion of chemistry in relation to the Earth, including life, although many aspects are treated too superficially for this to be a serious reference. A more detailed and professional treatment of the inorganic chemistry of the Earth is that of Henderson (1982). McElhinney (1976) contains a useful selection of articles on all aspects of the Earth's history, including its chemistry.

For readers interested in the detailed structure and properties of minerals themselves, Deer *et al.* (1966) can be recommended.

There is an enormous literature on more specialized aspects of geology and the Earth's chemistry. Books which have been of particular use in writing this chapter include Taylor and McLennan (1985) on the composition of the crust, Evans (1987) on ore-forming processes, and Bruland (1983) on the composition of the oceans. The articles by Meyer (1985) on historical aspects of ore formation, by Brimhall (1987) on patterns of elemental fractionation, and Li (1981, 1982) on oceanic residence times, are also interesting.

The use of inorganic elements in life is described by Phipps (1975).

6
Isotope distribution

Even before the end of the nineteenth century, it had been recognized that samples of lead from different sources had a slightly different relative atomic mass. With the discovery of isotopes (see Chapter 2), this phenomenon became explicable in terms of different proportions of the various stable isotopes. The isotopic compositions of elements, and the relative atomic masses which result from them, are normally regarded as fixed quantities. They are, indeed, determined primarily by factors of nuclear stability and particularly by the nucleosynthetic reactions which make the elements. As with the case of lead, however, detectable variations in isotopic composition have been found with many elements. These can be important in accurate analytical chemistry as they produce slight variations in relative atomic mass. They are also interesting for their own sake, because an understanding of their origin can give valuable information about the processes by which the elements have been formed and distributed.

As with the abundances of the elements themselves, isotopic compositions can be influenced by two factors. **Nuclear effects**, due to different rates of production and to radioactive decay of some isotopes are the most important. **Chemical differences** also exist, although they are much smaller than the differences between one element and another. In this chapter we shall look first at these different origins of isotopic variation, and then discuss the most important application of isotopic measurements, the dating of geological and other samples.

Nuclear processes affecting isotopic distribution

It is important, first of all, to note that the synthetic processes which have produced nuclides can lead to variations in relative isotopic abundances. This happens because most elements can be made by more than one route, with the different types of reactions producing isotopes of the same element in different relative yields. For example, the abundant nuclide ^{16}O is a major product of helium burning, whereas the rare ^{17}O is a byproduct of the CNO hydrogen burning cycle (see Fig. 3.3). ^{18}O may be made by both routes, although not necessarily under exactly the same conditions as those by which most ^{16}O or ^{17}O is produced. Stars of different masses, where these reactions proceed to different extents, may therefore produce oxygen with a different isotopic composition. Supernovae probably give mostly ^{16}O, with very little of the other isotopes. These may be present in higher proportions in gas blown off from the surfaces of stars during or after the operation of the CNO reactions. Thus, the isotopic composition of interstellar gas may vary, according which types of star have produced the elements in a particular region. With this in mind, it is interesting to ask whether the solar system and the Earth itself were made of isotopically homogeneous material. In fact, there is strong evidence that, for the solar system as a whole, this is *not* the case. Studies of oxygen isotope distribution in some meteorites suggest that they were made from varying proportions of material from at least two sources, with a different isotopic composition. The nature of this evidence is discussed below, but the same kind of data suggests that the Earth itself was originally uniform in its isotopic composition. Possibly, this was not true of the material from which the Earth was made, but is a result of the heating processes which led to a thorough mixing of the different components at an early stage. Whatever the reason, it seems that isotopic variations on the Earth have mostly come about since its origin.

The most obvious nuclear process which leads to isotopic variation is radioactive decay. Several elements have long-lived radioactive isotopes, a selection of which was listed in Table 2.1. There is strong evidence that other, shorter-lived nuclides, including ^{26}Al, ^{129}I, and ^{244}Pu were originally present on the Earth. The abundance of all these species has clearly varied in time: the geologically important heat output from several radioactive nuclides was shown in Fig. 5.3, illustrating the decline in output from each species which follows its decay. The initial abundance of ^{40}K, for example, must have been some ten times its present value, while that of ^{235}U was nearly 100 times greater than at present. With this decline, the concentration of **radiogenic** nuclides—the stable products of radioactive decay—has increased. Most ^{40}Ar present on Earth comes from decay of ^{40}K, and ^{4}He is a product of α decay of heavy elements such as U and Th. The most complex element in this respect is lead; of the four stable isotopes, only one (^{204}Pb) is not radiogenic. ^{206}Pb forms the end of the decay series starting from ^{238}U (shown in Fig. 2.13 on p. 59), and in a similar way ^{207}Pb comes from ^{235}U, and ^{208}Pb from ^{228}Th. The proportion of all these radiogenic isotopes has been increasing since the formation of the Earth, at rates depending on the different half-lives of the precursor nuclides.

Variations of isotopic composition in *time* give rise to variations in *space* because of the different chemical compositions or rocks. For example, lead extracted from minerals containing uranium will be enriched in ^{206}Pb and ^{207}Pb, relative to lead from sulphide minerals which have very little uranium. In a similar way, xenon obtained from oil wells and originating deep in the crust has been found to be slightly enriched in ^{128}Xe, thought to be derived from the spontaneous fission of the now extinct ^{244}Pu. These variations of isotopic composition can give very important geological information and, indeed, they are widely used for dating rocks, as will be described later. As already mentioned, the variations in lead isotopes can give rise to quite significant variations in atomic mass, which were recognized before the radioactive origin of Pb was understood.

Although radioactive decay is the predominant nuclear process on Earth, there are other phenomena which can have a minor influence on isotope distribution. Nuclear reactions are produced by cosmic rays when they enter the upper atmosphere and, to a lesser extent, in surface rocks at ground level. Very small amounts of some radioactive nuclides are made by these reactions. The best known are ^{3}H and ^{14}C, produced when ^{14}N captures neutrons formed as spallation products:

$$n + {}^{14}N \begin{cases} \nearrow {}^{14}C + {}^{1}H \\ \searrow {}^{3}H + {}^{12}C \end{cases} \tag{6.1}$$

^{14}C has a half-life of 5730 years, and is present in a very small, nearly steady-state concentration, in the atmosphere (less than 10^{-6} per cent of all C) and in living organisms which derive their carbon content ultimately from atmospheric CO_2. Equilibration with atmospheric carbon ceases when an organism dies and so the ^{14}C content starts to decay. This forms the basis for **radiocarbon dating**. The development in recent years of highly sensitive mass spectrometers based on tandem accelerators led to the detection of a number of other radioactive species produced by cosmic ray bombardment. Some of these nuclides, which can also be used for dating purposes, are listed in Table 6.1.

In recent decades a wide range of radioactive nuclides, including isotopes of previously unknown elements, have been manufactured artificially. The ways in which this is done are described briefly in Chapter 2. Although the quantities manufactured make a negligible contribution to overall elemental abundances, they are significant as the only terrestrial source of many radioactive isotopes. Most come from the neutron-induced fission of ^{235}U in reactors and nuclear explosions, although many other species are also generated by neutron capture processes at the same time. Figure 6.1 shows the mass distribution of the fission products from ^{235}U, and illustrates how this depends on the

Table 6.1
Some radioactive nuclides produced by cosmic ray bombardment

Nuclide	*Half-life (years)*
^{10}Be	1.5×10^6
^{14}C	5730
^{26}Al	7.2×10^5
^{32}Si	276
^{36}Cl	3.1×10^5
^{39}Ar	269
^{53}Mn	3.7×10^6
^{59}Ni	8×10^4
^{81}Kr	2.1×10^5

Source: Faure (1986)

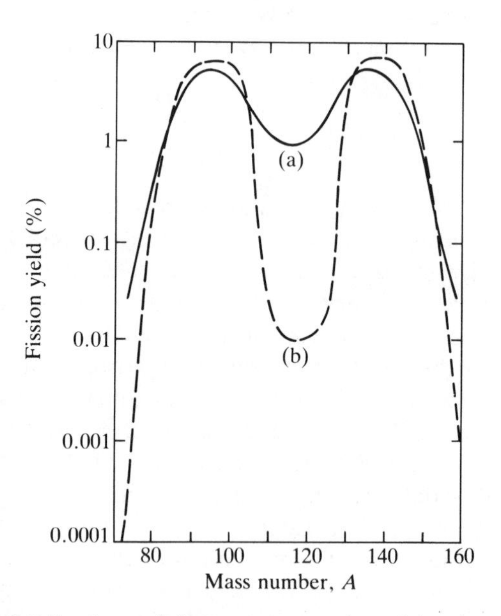

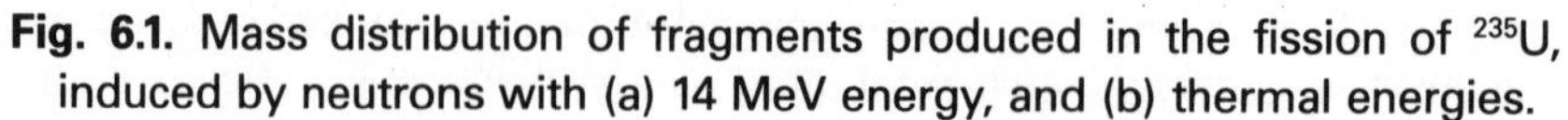
Fig. 6.1. Mass distribution of fragments produced in the fission of ^{235}U, induced by neutrons with (a) 14 MeV energy, and (b) thermal energies.

energies of the neutrons used. In controlled nuclear reactors, high-energy neutrons coming from fission events are **moderated**, or reduced to thermal energies, before initiating further fission. These low neutron energies lead predominantly to 'asymmetric' fission, with most products being concentrated in a **light fraction** from 90–105 mass units, and a **heavy fraction** with atomic masses between 130 and 145. The principal stable nuclides in these ranges are isotopes of elements between Y and Ru for the light fraction, and between Xe and Nd for the heavy fraction, but most of the immedidate products of fission are radioactive nuclei on the neutron-rich side of the line of stability, which undergo β decay. Significant amounts of some radioactive isotopes have been released into the environment during the atmospheric testing of nuclear weapons in the 1950s, from the fire at the nuclear reprocessing plant at Windscale in Britain in 1956, and more recently, from the fire in a nuclear power plant at Chernobyl in the Ukraine in 1987. Radiation received by human populations comes mostly from isotopes of elements which are taken up by the body, generally from the soil via food of different kinds. These isotopes include ^{131}I and ^{133}I, with half-lives of 8 days and 21 hours, respectively, and so important on a short time-scale, as well as longer-lived ones such as ^{90}Sr (half-life 28 years) and ^{137}Cs (half-life 29 years).

Nuclear reactors operate under highly artificial conditions and it is, therefore, surprising to find that a fission chain reaction occured naturally in at least one deposit of uranium ore. The initial evidence for this came from the isotopic composition of uranium extracted from the Oklo mine in Gabon, West Africa. The ^{235}U content is significantly less than the normal 0.72 per cent and, indeed, in the most concentrated parts of the deposit is as low as 0.44 per cent. The only reasonable cause for such a strong depletion is that a proportion of the ^{235}U was consumed by a chain reaction, sometime after the formation of the deposit. This explanation is supported by the occurrence of fission products in the deposit. Figure 6.2, for example, shows how the isotopic composition of the lanthanide element Nd found in the Oklo ore compares with the normal natural abundance, and with that expected from ^{235}U fission. The Oklo Nd deviates strikingly from the normal abundance pattern. Although it is rather different from the expected fission product abundance, this can be accounted for, firstly, by assuming that a small amount of natural Nd was also present and, secondly, by allowing for neutron capture processes during the reaction, the main effect of which was to convert some ^{143}Nd into ^{144}Nd.

The chemical features of the Oklo deposit nicely exemplify some of the ideas which have been discussed in previous chapters. In the first place, the fission products remaining in the deposit are isotopes of elements which are generally quite insoluble in aqueous environments. These include Y and lanthanides, Zr and Nb (all lithophiles with ionic charges between 3+ and 5+), and the siderophiles Ru–Ag. On the other hand, the much more soluble lithophilic elements Rb, Cs, Sr, and Ba (with 1+ and 2+ ionic charges) have all been washed out and no trace

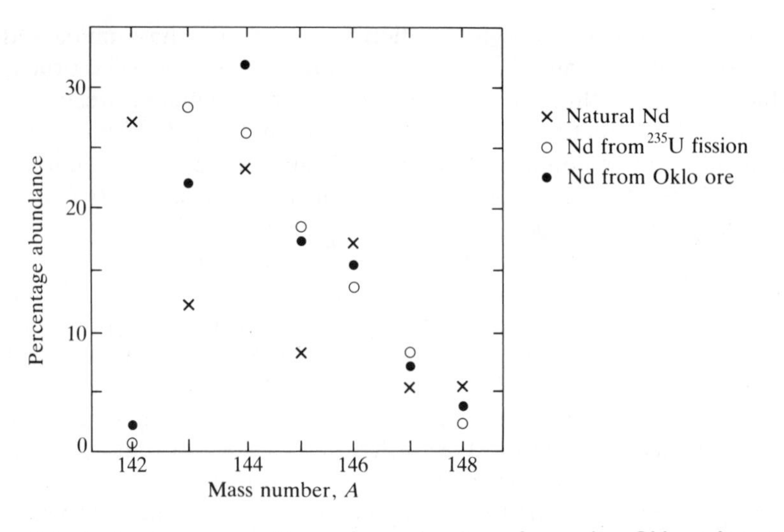

Fig. 6.2. Isotopic composition of neodymium from the Oklo mine and other sources. The Oklo sample is quite unlike most natural Nd. Its composition is close to that predicted from ^{235}U fission, with small modifications arising from neutron capture by ^{143}Nd and ^{145}Nd during the fission reaction.

of these fission products remain. Secondly, the formation of the deposit itself is quite interesting. In the early atmosphere the O_2 concentration was very low, and the stable form of uranium under these conditions was the very insoluble UO_2. It is likely that this mineral formed placer deposits in streams, in the same way that some heavy insoluble minerals do today. About 1800 million years ago, the development of photosynthesis led to a rapid increase in atmospheric O_2. Surface uranium deposits would then be oxidized to the much more soluble UO_2^{2+} complex ion and were, thus, washed down by rivers. The soluble uranium was then precipitated in sediments as UO_2, by reduction under anoxic conditions produced by bacteria and decaying organic matter. In this way, the highly concentrated Oklo ore was formed. The water present in the deposit acted as a moderator, thus assisting the development of the chain reaction. Yet even in pure UO_2, it seems such a reaction could not occur with the present ^{235}U abundance of 0.72 per cent. At the time the deposit was formed, however, the natural abundance would have been considerably higher, around 3 per cent. This would have been sufficient for a chain reaction to start and it continued to operate for some thousands of years, until about half the ^{235}U was burnt out in the most concentrated part. Although this so-called

Oklo phenomenon could not recur with the low ^{235}U abundance today, there is no reason why the conditions in the Oklo deposit should have been unique at the time of its formation. No strong evidence exists, however, that other uranium deposits have undergone a similar reaction; very small variations in the ^{235}U content of some uranium ores are found, but these might have resulted from processes of chemical fractionation which will now be described.

Chemical fractionation of isotopes

Chemists are usually able to ignore the fact that many of their supposedly 'pure' substances are actually isotopic mixtures. This is because the chemical behaviour of an element is determined primarily by the nuclear charge: that is, the atomic number of an element, which gives its position in the periodic table. However, the mass of an atom does have a small influence on its physical and chemical properties. Differences between isotopes are most obvious in spectroscopic measurements of the rotational and vibrational energy levels of molecules, but isotopic mass also has effects on both the rates and equilibrium constants of chemical reactions. These have caused some natural fractionation of isotopes in space and on the Earth, and are also exploited for the deliberate separation of isotopes both at a laboratory and an industrial scale. The requirements of the nuclear industry have led to changes in the isotopic composition of commercially available samples of a number of elements. Thus, ^{235}U separation, for enrichment of nuclear fuel elements or for weapons, means that most uranium sold for research purposes is strongly depleted in this isotope. Lithium from some sources is similarly depleted in ^{6}Li, used in the production of thermonuclear warheads.

The simplest isotopic effect is that which arises in the kinetic theory of gases. In a gas at a temperature, T, the mean kinetic velocity of a molecule with mass, M, is:

$$\bar{c} = (8kT/\pi M)^{1/2} \tag{6.2}$$

Thus, isotopes with different mass M will have different velocity distributions, which, in turn, produce differences in a number of properties. For example, the rate of **diffusion** of gases through small pores is proportional to $\bar{c}$, and lighter isotopes will diffuse faster. The largest difference between naturally occuring isotopes is a factor of $\sqrt{2}$ between H_2 and D_2. Even hydrogen isotopes could not be separated efficiently in a single stage, but gas diffusion plants for isotopic separation are based on this effect, amplified many times by performing the separation in a large number of sequential stages. The most important practical use of gas diffusion has been with the volatile compound UF_6 for the preparation of ^{235}U enriched uranium. Clearly, the separation factor of a single step is very small in this case, and many thousands of stages are necessary to give high levels of enrichment. The method, in fact, has now been largely superseded by one based on the ultracentrifuge. This also depends on gas kinetic factors, which cause the heavier isotope to concentrate towards the outer edge of the centrifuge. More efficient separation can also be obtained by combining diffusion with a thermal gradient, and such a method is used sometimes on a laboratory scale for preparing small amounts of pure isotopes for research purposes.

As well as being useful for the artificial separation of isotopes, gas kinetic differences may have played a role in natural processes. For example, lighter isotopes will be lost more readily from the atmosphere of a planet. The loss of hydrogen, through the photodissociation of water in the early atmosphere, has given rise to a considerable deuterium enrichment on Venus, and may have operated to a smaller extent on the Earth. Using the same idea, the high $^{15}N/^{14}N$ ratio in the atmosphere of Mars has been used to estimate how much nitrogen has been lost since its formation (see Chapter 4).

Other chemical differences between isotopes are more subtle and depend on quantum mechanical effects. They arise because the allowed energy levels for a molecule depend on the masses of the atoms. The major effect comes from the so-called **zero-point energy** of vibration. Even at very low temperatures, the atoms in a molecule or solid are not stationary, but have small random motions governed by the uncertainty principle of quantum mechanics. Solution of Schrodinger's equation shows that the lowest allowed vibrational energy is given by:

$$E_0 = 1/2(h/2\pi)(k/m)^{1/2} \qquad (6.3)$$

where h is Planck's constant, k is the bond force constant (the same for different isotopes), and m is the mass of the vibrating atom. For hydrogen and deuterium the masses differ by a factor of 2 and, thus, the zero-point energy is a factor of $\sqrt{2}$ lower for deuterium. When a molecule is dissociated into atoms the zero-point energy disappears, and so the difference in this energy for two isotopes gives rise to slightly different bond dissociation energies (see Fig. 6.3). For example, DCl has a dissociation energy about 4.8 kJ/mol larger than that of HCl. Similar differences arise with other isotopes, although as with gas kinetic properties, these are generally much smaller than between H and D.

The above discussion shows that chemical bonds are more stable with

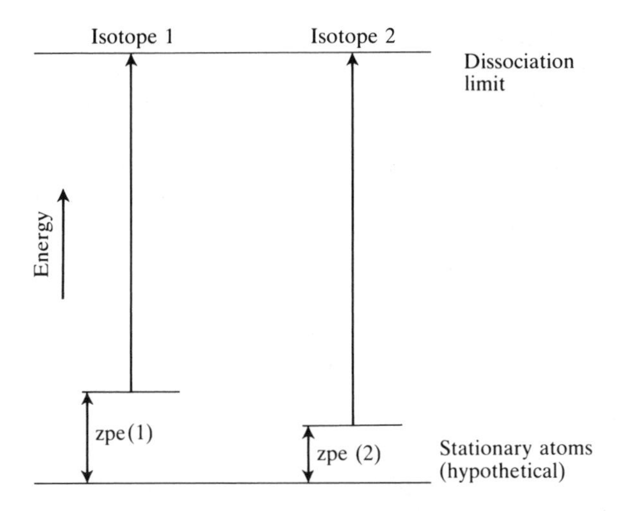

Fig. 6.3. If nuclei were stationary in the ground state of a molecule, different isotopes would have the same bond dissociation energy. Differences in the dissociation energy occur because of the zero-point vibrational energy (zpe), explained in the text.

heavier isotopes of a given element than with lighter ones. This difference can influence both reaction rates and equilibria. The **kinetic isotopic effect** arises when the energy barrier in a chemical reaction depends on the energy necessary to break a bond. For example, when an organic reaction has a mechanism involving C–H bond breaking as a rate-determining step, it will become slower if the H is replaced by D. The same effect operates in the electrolysis of water, where the gas evolved has relatively more H, and the water itself becomes enriched in D. The influence on chemical equilibria may not be so pronounced, because it will generally depend on the *differences* of zero-point energy between different molecules. Nevertheless, the reaction

$$H_2O(l.) + HD(g.) = HDO(l.) + H_2(g.) \qquad (6.4)$$

has an equilibrium constant $K = 4.5$ at 0°C. This equilibrium is certainly a major factor contributing to the abundance of deuterium on earth, where the D/H ratio is nearly ten times that of the solar system as a whole.

Although the effects are largest with H and D, isotopic differences due to chemical fractionation have been detected in a number of elements, and some of these are used by geologists to give information about the origin of rocks. For example, different minerals have slightly different equilibrium constants of incorporating ^{18}O and ^{16}O. These vary with temperature in a way that can be predicted theoretically or measured in the laboratory, and are illustrated for three different mineral-water equilibria in Fig. 6.4. Thus, the measured $^{18}O/^{16}O$ ratio can give information about the temperature at which a mineral was formed. The first application of this method was to sedimentary carbonate deposits, utilizing the isotopic fractionation between CO_2 and CO_3^{2-}. From the known temperature variation of this equilibrium, an attempt was made to deduce the mean ocean temperatures existing at times when the carbonate deposits were laid down. The causes of isotopic variation in carbonate deposits are now thought to be more complicated, as it appears that the ^{18}O content of sea water is itself subject to variation, as a result of fractionation with atmospheric oxygen. Nevertheless, oxygen isotope data form an important source of information on climatic changes in the past, and particularly on the sequence of ice ages which have occurred during the past million years or so. The method has also been used to investigate various aspects of mineral formation. Another useful isotope pair is $^{34}S/^{32}S$, which has been applied to the study of the formation of sulphide ores by hydrothermal processes.

It is normally assumed in the studies just described that chemical fractionation is the *only* source of isotopic variation. Clearly, it is

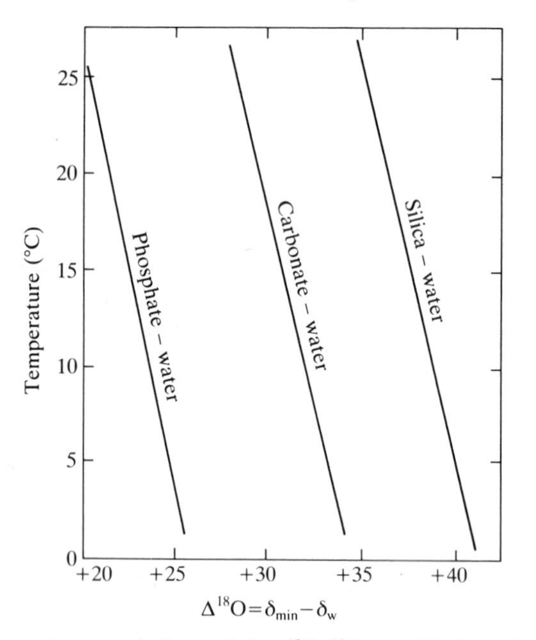

Fig. 6.4. Temperature variation of the $^{18}O/^{16}O$ equilibrium between water and different minerals. The values are deviations in the $[^{18}O]/[^{16}O]$ ratio, measured in parts per thousand. (From Faure 1986.)

essential for this that none of the isotopes concerned has a radiogenic origin. Another implicit assumption, however, is that the Earth was originally formed with a uniform isotopic composition. It is actually possible to test this assumption for an element such as oxygen which has more than two stable non-radiogenic isotopes. The chemical difference between isotopes (except very light ones such as H and D) is very nearly proportional to the difference in mass. It follows that any chemical process which alters the relative $^{18}O/^{16}O$ ratio should affect the relative $^{17}O/^{16}O$ ratio by just about *half* as much. If one makes a plot of these two deviations, for samples fractionated to different extents from material with the same original isotopic composition, the points should fall on a straight line, with a slope of 0.5. Such a plot, showing oxygen samples derived from different sources, is shown in Fig. 6.5. All terrestrial samples fall on the line (a), which has the correct slope for chemical fractionation. It has also been found that samples from lunar rocks fall on the same line. This is strong evidence that all these rocks contain oxygen from the same original isotopic mixture. However, samples from many meteorites do not follow the same pattern. In fact, different mineral inclusions found in the *same* carbonaceous chondrite meteorite

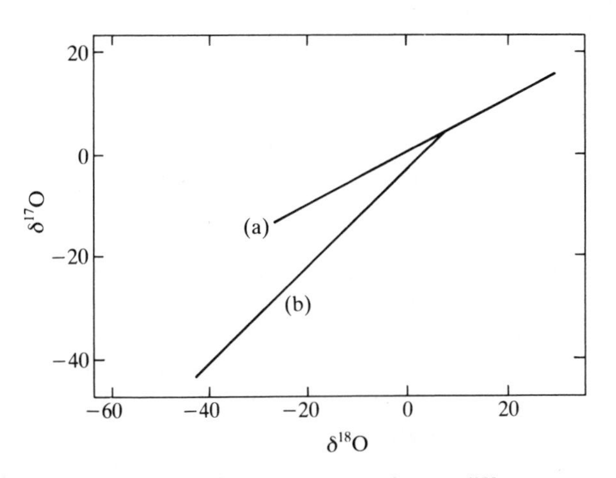

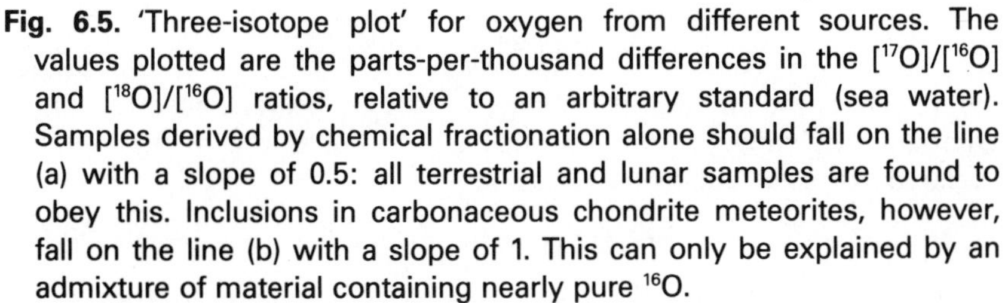

Fig. 6.5. 'Three-isotope plot' for oxygen from different sources. The values plotted are the parts-per-thousand differences in the $[^{17}O]/[^{16}O]$ and $[^{18}O]/[^{16}O]$ ratios, relative to an arbitrary standard (sea water). Samples derived by chemical fractionation alone should fall on the line (a) with a slope of 0.5: all terrestrial and lunar samples are found to obey this. Inclusions in carbonaceous chondrite meteorites, however, fall on the line (b) with a slope of 1. This can only be explained by an admixture of material containing nearly pure ^{16}O.

often fall on a line of unit slope (labelled (b) in Fig. 6.5). No known chemical or nuclear process occuring within the solar system can account for this. The only reasonable explanation is that the meteorites were made from varying proportions of material from two distinct sources, with different isotopic compositions. The fact that line (b) has a the slope close to one indicates an admixture of almost pure ^{16}O. It is thought that this must have come from a supernova, probably the same one which generated ^{26}Al and other short-lived isotopes. Isotopic anomalies have also been found with some other elements and all this evidence lends support to the view that the solar system was formed from a rather heterogeneous mixture of gases from different sources. Such isotopic heterogeneity also suggests that the early solar system may have been chemically inhomogeneous. As discussed in Chapter 4, this is one of the problems faced by theories of the origin and the chemical composition of the planets.

Isotope dating methods

The steady decay of radioactive nuclides, at a rate which is, in most cases, completely insensitive to conditions of pressure, temperature or chemical combination, forms an ideal 'clock' for providing absolute ages of rocks and other materials. Indeed, all the precise information that Earth scientists have accumulated about the age of the solar system and the rocks in the Earth's crust comes from this source. A wide variety of isotopes is available, including all the long-lived species listed in Table 2.1, and all the shorter-lived nuclides derived from cosmic ray bombardment, which were listed in Table 6.1.

One of the best known dating procedure is the **radiocarbon method**, based on the decay of ^{14}C, with a half-life of 5730 years. As explained above, this is present as a very small, but nearly constant fraction of the carbon in all living organisms. When a plant or animal dies the ^{14}C decays, and the amount remaining can thus be used to determine the time since death. The basic radioactive decay law used for this, and for other methods, is:

$$[A]=[A]_0 e^{-\lambda t}. \tag{6.5}$$

Here $[A]$ is the amount of radioactive isotope present now, $[A]_0$ is the amount originally present t years ago, and λ is the **decay constant**, related to the half-life by the equation:

$$t_{1/2}=\ln 2/\lambda \tag{6.6}$$

The time scale of ^{14}C decay makes this method appropriate for archaeological specimens, with ages between a few hundred and a few tens of thousands of years. It can clearly be applied to dead plant matter, animal or human bones, or to any artifacts made from them. In the traditional method, organic matter containing a known amount of carbon is converted to CO_2, and the radioactivity determined in a special gas counter. This is necessary because of the very low energy of the β particles emitted by ^{14}C, which do not pass through the window of a normal Geiger counter. Recently, much more accurate measurements of the ^{14}C content have been possible using mass spectrometers.

To obtain radiocarbon dates, it is clearly necessary to know the original ^{14}C concentration in the sample. In early work it was assumed that the atmospheric content had not changed with time, so that the ^{14}C abundance measured today could be used for the starting point. However, evidence accumulated that the resulting radiocarbon dates were sometimes quite inaccurate, and the assumption of constant ^{14}C abundance has been abandoned. It has therefore been necessary to calibrate the technique on samples for which the age is known independently. The most useful calibration method has been based on **dendochronology**, or 'tree-ring dating'. This depends on observing the pattern of annual growth rings, which show characteristic variations due to fluctuations in the climate. By matching up the patterns from series of specimens, starting with trees still living and working back into the past, it has been possible to obtain an absolute age calibration for the last 70 thousand years or so. Thus, information can be obtained about the variation in atmospheric ^{14}C content over that period. It is clear from these results that the proportion of ^{14}C has declined markedly in the last century. This is attributed to the general increase of atmospheric CO_2 from burning fossil fuels, which have almost no ^{14}C in them. Variations in the past are less easy to explain. Climatic changes may have altered the general level of CO_2 in the atmosphere, but it is also possible that variations in solar activity may have led to changes in the rate of ^{14}C production in the upper atmosphere. Whatever their origin, however, these changes are now known well enough to give much more reliable ^{14}C dates.

A minor problem in radiocarbon dating is the possibility of chemical fractionation, which leads to a ^{14}C concentration in the living organism slightly different from the atmospheric value. Fortunately, this can be allowed for by also measuring the $^{13}C/^{12}C$ ratio, which can be altered only by chemical effects. As with the oxygen isotope fractionation discussed above, chemical variation in the relative $^{14}C/^{12}C$ ratio should be nearly double to that for $^{13}C/^{12}C$.

The other nuclides listed in Table 6.1 have been used in a similar way to ^{14}C, for dating geological processes, such as sedimentation, which take place over a time scale of up to a few million years. An example of this given in Fig. 6.6, which shows the variation with depth of the ^{10}Be and ^{26}Al contents in a manganese module. Like ^{14}C, these nuclides are present in the ocean in nearly constant steady-state concentrations and are, thus, incorporated into the surface of the growing nodule. The decline in concentration within the nodule comes from their radioactive decay and allows the rate of growth to be determined. Both the ^{10}B and ^{26}Al data shown are consistent with a growth of about 2.5 mm per million years.

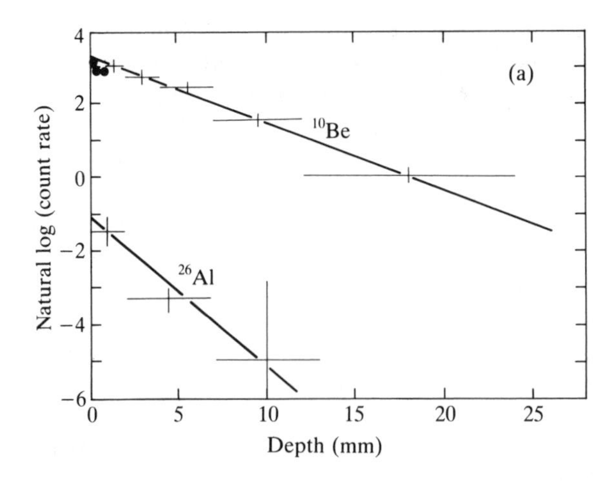

Fig. 6.6. The variation with depth of ^{10}Be and ^{26}Al content (measured as the radioactive count rate) in a manganese nodule from the Pacific Ocean. The slopes of the lines are consistent with deposition rate of about 2.5 mm per million years. (From Faure 1986.)

Geological dating over longer time scales clearly relies on nuclides with appropriate half-lives. The naturally occurring species shown in Table 2.1 are suitable for this, and can provide dates in the range from tens to thousands of millions of years. The principles involved are best illustrated by one of the most widely used and straightforward methods, the Rb/Sr technique. Naturally, occurring rubidium contains 28 per cent of the radioactive isotope ^{87}Rb, which undergoes β decay to ^{87}Sr with a half-life of 4.9×10^{10} years. The amount of ^{87}Rb remaining after t years can be expressed in terms of its original concentration $[^{87}Rb]_0$ by equation 6.5:

$$[^{87}\text{Rb}] = [^{87}\text{Rb}]_0 e^{-\lambda t} \qquad (6.5')$$

The ^{87}Sr concentration will grow at the same rate, so that:

$$[^{87}\mathrm{Sr}] = [^{87}\mathrm{Sr}]_0 + [^{87}\mathrm{Rb}]_0(1 - e^{-\lambda t}) \tag{6.7}$$

This equation can be put in a more useful form, firstly by using equation 6.5 to express it in terms of the *present*, rather than the *initial* ^{87}Rb concentration, and secondly, by dividing by the concentration of ^{86}Sr:

$$[^{87}\mathrm{Sr}/^{86}\mathrm{Sr}] = [^{87}\mathrm{Sr}/^{86}\mathrm{Sr}]_0 + [^{87}\mathrm{Rb}/^{86}\mathrm{Sr}](e^{\lambda t} - 1) \tag{6.8}$$

It is generally easier to measure ratios of isotopic concentrations than absolute values, and ^{86}Sr is an ideal 'internal standard', as it is neither radioactive nor radiogenic, so that its concentration should not change with time. The age of a sample could be found from the above equation, if the present [^{87}Sr/^{86}Sr] and [^{87}Rb/^{86}Sr] ratios were measured, and the initial ratio $[^{87}\mathrm{Sr}/^{86}\mathrm{Sr}]_0$ were known. Unfortunately, there is no direct way of estimating the initial ratio, but the problem can be overcome by measuring the present isotopic ratios in a number of minerals formed at the same time. When magma solidifies, the various minerals crystallize with different relative concentrations of rubidium and strontium. For example, potassium feldspar has a relatively high concentration of rubidium, because of the chemical similarities of K and Rb; on the other hand, calcium-rich minerals will tend to concentate strontium. Thus, a given rock will contain minerals will contain different [^{87}Rb/^{86}Sr] ratios, formed at the same time. Since chemical fractionation between the fairly heavy atoms ^{87}Sr and ^{86}Sr should be negligible, it can be assumed that the initial ratio of these isotopes was the same in all the minerals present. The technique, therefore, consists of plotting the [^{87}Sr/^{86}Sr] ratio of each mineral against its [^{87}Rb/^{86}Sr] ratio, and if the above assumptions are correct, equation 6.8 shows that a straight line should be obtained, with a slope of $(e^{\lambda t} - 1)$. This is illustrated in Fig. 6.7, showing data from a rock sample from eastern Greenland. The slope of the line shows that this rock solidified 3.66 billion years ago, and represents one of the earliest rocks found on the crust of the Earth. The same method has been applied to meteorites and rocks from the Moon, as well as a wide range of terrestrial samples. An age of 4.6 billion years has been found for many meteorites and for rocks from the lunar highlands. This is generally assumed to be the age of the solar system, and is consistent with quite different, astrophysical, estimates of the Sun's age. On the other hand, most rocks in the Earth's crust are much younger, which is clear evidence of the importance of tectonic processes that form and recycle the crust.

The intercept of the line in Fig. 6.7 gives the value of $[^{87}\mathrm{Sr}/^{86}\mathrm{Sr}]_0$, which is the isotopic composition of the magma from which the rocks formed. Samples from meteorites have a value of 0.699 for this ratio,

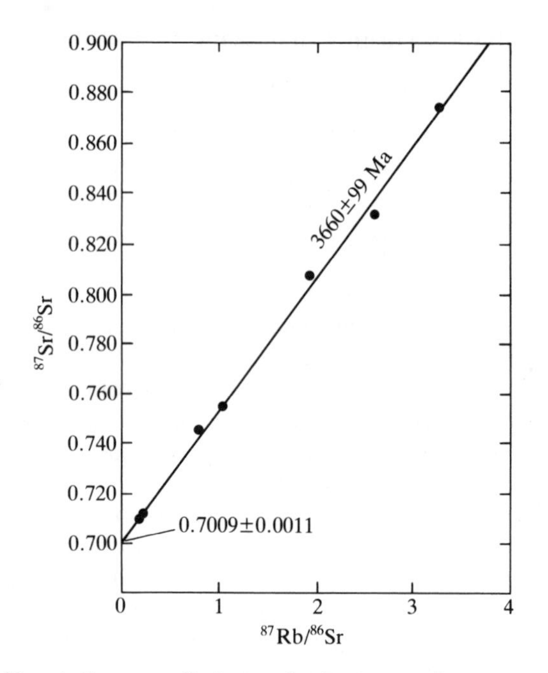

Fig. 6.7. Rb/Sr dating applied to Amitsoq gneiss, a rock found in Greenland. The slope of the line indicates an age of 3700 million years, representing one of the oldest rocks of the Earth's crust. (From Faure 1986.)

which must be the original composition of Sr in the solar system. With the slow decay of ^{87}Rb, however, the ratio has been increasing. For example, the present value in the mantle is 0.704. Most crustal rocks, however, contain a higher concentration of rubidium than the mantle, so that $[^{87}Sr/^{86}Sr]$ in the crust has been increasing at a faster rate. Measurement of the initial isotope ratio can, therefore, give important information about the origin of the magma from which rocks formed. Magma derived directly from the mantle will have a low value, certainly never greater than 0.704. Many granite samples show initial ratios higher than this, showing that they must be derived at least partly by the remelting of crustal rocks, which were differentiated from the mantle in some previous tectonic process.

Other isotope pairs are used in a way similar to the Rb/Sr method, to obtain different, or complementary information. These include K/Ar and Th/U/Pb methods. Often, the interpretation of these results is more difficult than for the Rb/Sr method. For example, loss of argon by diffusion from rocks can complicate the K/Ar method, although ways have been found of correcting for this problem. A frequently used method with lead is to look at the proportions of the two radiogenic isotopes ^{206}Pb and ^{207}Pb—derived at different rates from the decay of ^{238}U and ^{235}U, respectively—relative to the non-radiogenic ^{204}Pb. Again, the interpretation may be complicated by metamorphic processes which result in the loss or addition of lead to the rocks after they were originally formed. However, it is clear from these studies that the [$^{206}Pb/^{204}Pb$] and [$^{207}Pb/^{204}Pb$] ratios in terrestrial lead samples have evolved in a way which indicates that the Earth formed from material with the same original isotopic composition as that of meteorites, and at the same time, 4.6 billion years ago.

Study of the isotopic composition of meteorites has been used in a rather more ambitious way, to give information about the age of the elements in the solar system, and of the universe itself. Many samples from meteorites contain an excess of ^{26}Mg, which seems to be derived from the decay of ^{26}Al, with a half-life of 7×10^5 years. This idea is supported by the observation that the amount of ^{26}Mg excess is correlated with the Al content of the minerals. It is estimated from this that the solar system must contain some material which was released by a supernova only 2 million years or so before its formation. On the other hand, some other radiogenic isotopes present suggest a different age. For example, the isotopic composition of xenon in some meteorites shows excesses in ^{129}Xe, the decay product of ^{129}I, and ^{128}Xe formed in the spontaneous fission of ^{244}Pu. The parents involved here have much longer half-lives than ^{26}Al, and the amounts of the daughter nuclides present suggest that element synthesis took place some 200 million years before the solar system was formed. Like the oxygen isotope anomalies mentioned above, this shows that the solar system was formed from material from more than one source. The 2-million-year time scale is strong indication that a nearby supernova could have started the gravitational contraction which formed the solar system. The presence of older material suggests that star formation and element synthesis proceeds in cycles, which are thought to be related to the rotation of the spiral arms of the galaxy.

The methods by which the age of the universe has been estimated can be illustrated by the following, rather simplified argument. The present concentration of stable heavy nuclides is the result of the cumulative synthesis of the elements over the duration of the universe. Consider, on the other hand, a nuclide such as ^{235}U. The half-life (7×10^8 years) is considerably less than the age of universe, and so its concentration will reach a steady-state value, where the rates of production and decay are the same. Thus, the observed abundance of ^{235}U and its known decay constant can be used to estimate its rate of formation. If the relative rates of synthesis of different elements can be calculated from theory (based on neutron capture cross-sections as discussed in Chapter 3), then the rate of production of these other elements is also known. The cumulative concentration of a stable nuclide can, therefore, show how long the process has been occurring, at least if the rate of synthesis has not changed with time. In practice, the arguments used are slightly more sophisticated. By using a number of isotopic ratios, for example, $^{235}U/^{238}U$ and $^{228}Th/^{238}U$, it is possible to allow for some variation in the rate of synthesis since the universe began. There are many uncertainties in these estimates, but they are, nevertheless, consistent with a model in which the rate of nuclear synthesis has been decreasing exponentially since an origin between 16 and 20 billion years ago. This is similar to the age of the universe estimated by completely independent methods, based, for example, on the observed rate of expansion.

Summary

Small differences in the isotopic composition of elements, which give rise to deviations in relative atomic masses, can come from a variety of sources. Isotopes of many elements are not produced in the same proportions by stars of different types and, therefore, interstellar gas is not uniform in its isotopic composition. There is evidence that the solar system contains material from at least two sources, each with a different isotopic mixture. Isotopic variations on the Earth arise both from radioactive decay—of long-lived elements originally present and of shorter-lived species produced by cosmic ray bombardment—and from fractionation processes resulting from small chemical differences between isotopes.

Variations in isotopic composition are widely used in geological sciences. The existence on Earth of radioactive isotopes with a wide variety of half-lives enables the accurate dating of samples with ages ranging from hundreds to billions of years. Chemical fractionation of isotopes is sensitive to temperature, and can give information about past climates and the mechanism of mineral formation.

The age of the Earth, of the elements which make it up, and of the universe itself, have all been estimated by isotopic methods.

Further reading

Data on the isotopic compositions of elements, and on half-lives of both natural and artificial radioactive isotopes, can be found in any recent edition of the *CRC Handbook of Chemistry and Physics.* An authoritative review and tabulation of isotopic composition is given in IUPAC (1983*b*), from which Appendix B in this book has been compiled.

The application of isotopes to dating and other problems in geology is discussed in many books. There are chapters in Mason and Moore (1982), and Henderson (1982); Faure (1986) gives an excellent, more detailed account of the subject.

Evidence for the uneven isotopic composition of the early solar system is discussed by Clayton (1978), and Fowler (1984) describes how the age of the universe can be estimated from isotope abundances.

Appendix A
Elemental abundances

Table A.1 gives the values for the elemental abundances used in constructing the figures in the text. The difficulties in obtaining 'reliable' values have been referred to at various points in the main text, and it is important to realize that many estimates have been published, which often disagree to some extent. Most such differences are more of detail than essential substance, however, and do not affect the main conclusions drawn in the text. For this reason a critical assessment of different estimates has not been attempted, and most of the tables below are based on relatively few sources. References are given, so that interested readers can look at the different values and make their own conclusions about reliability.

For convenience, all the values in the table are presented in logarithmic form: that is, as $\log_{10}(A/\text{units})$, where A is the abundance, and the units, which are explained below for each column separately, are chosen to avoid negative numbers as far as possible. A blank means that no estimate is available, often—but *not* necessarily—because the relevant abundance is very low. The number of significant figures quoted for each numerical value is intended to give an idea of its likely accuracy: once again, the reader who is interested in more detailed error bounds can consult the literature.

Column a gives solar (spectroscopic) abundances from Ross and Aller (1976). The units are **numbers of atoms per 10^{12} of hydrogen**.

Column b shows abundances found in type C1 carbonaceous chondrite meteorites, also as **relative numbers of atoms**, normalized so as to correspond as closely as possible to the solar values for abundant and involatile elements such as Mg, Si, and Fe. The basic sources are Mason (1971) and Cameron (1973), but later corrections have been taken from Anders and Ebihara (1982), who give detailed references to analyses of individual elements.

Column c is an estimate of the overall composition of the Earth, expressed as **mass fraction in parts per billion**: that is μg/kg. This is undoubtedly the most 'problematic' of the abundance tables given here. The basic data are taken from Smith (1977), who obtained them by summing estimates for crust, mantle, and core compositions. However, a number of corrections have been made. In the first place, abundances of many lithophilic elements have been adjusted to take into account more recent estimates of the composition of the mantle (Jacoutz *et al.* 1979; Wanke 1983; Taylor and McLennan 1985). The following additional corrections have also been made (in $\log_{10}$ units): P (+0.5), Ga (−1.0), Ge (−0.5), Cd (−1.7), Hg (−0.6), Pb (−0.8), and Bi (−1.0). These are intended to correct the core composition assumed by Smith (1977). Although the adjustments are rather *ad hoc* in nature, they bring the relative abundances, both with respect to solar system and crustal values, more into line with elements of similar chemical behaviour.

Other discussions of the elemental composition of the Earth are given by Ganapathy and Anders (1974), and by Henderson (1982).

Column d shows an estimate of the average composition of the continental crust. The units are the same as in **c**: μg/kg. Data for most metallic elements are from Taylor and McLennan (1985), who give a detailed discussion of the methods used to obtain their estimates. Values for H, C, N, O, S, and halogens are those quoted by Smith (1977), and mostly obtained from Ronov and Yaroshevsky (1969). Rough estimates for noble gases, Te, Pt group metals, Ra, and Pa are from Mason and Moore (1982).

Column e shows the composition of dissolved species in sea water of average salinity (35 parts per thousand). Most data are from a compilation by Bruland (1983), except for noble gases and radioactive elements, which are taken from Henderson (1982). The unit is **10^{-12} mol/kg water**, that is **picomolal concentration.**

Column f gives the elemental composition of the human body, based on Bowen (1979). The abundances (as is normal for biological specimens) are expressed as a proportion of the dry mass. The unit is **mass fraction in parts per million**, or mg/kg dry mass.

Table A.1

Abundances of the elements

Element	*Log_{10} (relative abundance)*					
	a *Sun*	*b* *Meteorite*	*c* *Earth*	*d* *Crust*	*e* *Ocean*	*f* *Body*
H	12.00	8.4	4.8	6.2	—	6.8
He	10.8	—	−3	0	3	—
Li	1.0	3.4	3.2	4.1	7.41	0
Be	1.2	1.4	2	3.2	1.3	−1
B	<2	3.0	2.6	4.0	8.62	2

Table A.1 (*cont.*)

Element	*Log$_{10}$ (relative abundance)*					
	a *Sun*	*b* *Meteorite*	*c* *Earth*	*d* *Crust*	*e* *Ocean*	*f* *Body*
C	8.6	7.4	5.1	6.7	9.3	7.2
N	7.9	6.3	4.3	4.3	7.5	6.3
O	8.8	8.47	8.47	8.67	—	7.6
F	4.6	4.5	4.2	5.8	7.83	3.4
Ne	7.6	—	—	0	4	—
Na	6.3	6.36	6.3	7.36	11.67	5.0
Mg	7.6	7.63	8.18	7.51	10.73	4.3
Al	6.5	6.53	7.17	7.92	4.3	2
Si	7.7	7.60	8.16	8.43	8	4
P	5.5	5.61	6.2	6.0	6.3	5.9
S	7.2	7.3	7.64	5.8	10.45	5.2
Cl	5.5	5.3	3.8	5.3	11.74	5.0
Ar	6.0	—	1	1	7.0	—
K	5.2	5.17	5.1	6.96	10.01	5.1
Ca	6.4	6.38	7.23	7.72	10.01	6.0
Sc	3.0	3.1	4.1	4.5	1.2	—
Ti	5.1	5.0	5.97	6.7	4	—
V	4.0	4.1	4.7	5.4	4.5	−1
Cr	5.7	5.72	6.5	5.3	3.6	0
Mn	5.4	5.79	5.86	6.1	2.7	1
Fe	7.5	7.55	8.48	7.84	3	3.6
Co	4.9	4.95	6.02	4.5	1	0
Ni	6.3	6.29	7.21	5.0	3.9	0
Cu	4.1	4.3	5.07	4.9	3.6	2
Zn	4.5	4.70	4.6	4.9	3.8	3.3
Ga	2.8	3.2	3.3	4.3	2	—
Ge	3.5	3.67	4.3	3.2	2	—
As	—	2.4	3.4	3.4	4.3	1
Se	—	3.4	4.1	1.7	3	0
Br	—	2.4	1.5	3.4	8.92	2.4
Kr	—	—	—	−1	3	—
Rb	2.6	2.5	2.7	4.5	6.2	2.8
Sr	2.9	2.97	4.2	5.4	8.0	2.4
Y	2.1	2.3	3.4	4.3	2	—
Zr	2.8	2.6	3.6	5.0	2	0

Table A.1 (*cont.*)

Element	*Log$_{10}$ (relative abundance)*					
	a *Sun*	*b* *Meteorite*	*c* *Earth*	*d* *Crust*	*e* *Ocean*	*f* *Body*
Nb	1.9	1.5	2.8	4.0	1	—
Mo	2.2	2.0	3.3	3.0	5.0	1
Ru	1.8	1.9	3.1	0	—	—
Rh	1.4	1.1	2.5	−1	—	—
Pd	1.5	1.7	2.9	0	—	—
Ag	0.9	1.3	1.8	1.9	1	—
Cd	1.9	1.9	1.9	2.0	3	1.7
In	1.7	0.9	0.6	1.7	0	—
Sn	2.0	2.2	3.1	3.4	1	—
Sb	1.0	1.1	2.2	2.3	3	—
Te	—	2.3	2.7	−1	—	—
I	—	1.6	1.6	2.7	5.6	2
Xe	—	—	—	−2	3	—
Cs	<1.9	1.2	1.1	3.0	3.3	—
Ba	2.1	2.2	3.7	5.4	5	1
La	1.1	1.3	2.6	4.2	1.5	—
Ce	1.6	1.7	3.3	4.5	1.3	—
Pr	0.7	0.8	—	3.6	1	—
Nd	1.2	1.5	3.1	4.2	1.3	—
Sm	0.7	1.0	2.3	3.5	0.6	—
Eu	0.7	0.6	1.8	3.0	0.0	—
Gd	1.1	1.1	2.5	3.5	0.8	—
Tb	—	0.4	—	2.8	0	—
Dy	1.1	1.2	2.3	3.6	0.8	—
Ho	—	0.5	—	2.9	0	—
Er	0.8	1.0	2.0	3.3	0.7	—
Tm	0.3	0.2	—	2.5	0	—
Yb	0.9	1.0	2.4	3.3	0.7	—
Lu	0.8	0.2	1.6	2.5	0	—
Hf	0.8	0.8	2.6	3.5	1	—
Ta	—	0.0	1.4	3.0	1	—
W	1.7	0.7	2.5	3.0	2.7	—
Re	<−0.3	0.3	1.3	−0.3	1.3	—
Os	0.7	1.5	2.8	−1	—	—
Ir	0.85	1.4	2.6	−1	—	—

Table A.1 (*cont.*)

Element	*Log₁₀ (relative abundance)*					
	a *Sun*	*b* *Meteorite*	*c* *Earth*	*d* *Crust*	*e* *Ocean*	*f* *Body*
Pt	1.8	1.7	3.2	0	—	—
Au	0.8	0.9	2.5	0.5	1.4	—
Hg	<2.1	1.3	1.5	2.0	1	—
Tl	0.9	0.9	0.9	2.6	1.8	—
Pb	1.9	2.1	2.7	3.9	1	2
Bi	<1.9	0.8	1.1	1.8	−1	—
Ra	—	—	—	−4	−4	−7.5
Th	0.2	0.1	1.8	3.6	1.6	—
Pa	—	—	—	−5	−4	—
U	<0.6	−0.4	1.4	3.0	4.1	−1.0

Appendix B
Isotopic abundance and variation

Table B.1 shows the isotopic composition of the naturally occurring elements. The accompanying notes, explained below, show sources of variability. When there is a well-established isotopic variation in natural terrestrial sources of an element (**note v**), the extent of variability is normally at the level of the last significant figure quoted. In this case, the percentage abundances given are those likely to be found in laboratory samples, and not an overall average. Where no normal terrestrial variability has been established, the number of significant figures given reflect the accuracy of measurement. The table is adapted from IUPAC (1983*b*).

Notes

- v Elements for which isotopic **variability** has been established in normal samples of terrestrial origin.
- e Elements for which natural samples of **exceptional** provenance (for example meteorites or the Oklo mine) may have a composition outside the normal range.
- m Elements where some commercially available samples may have a **modified** composition, as a result of deliberate or inadvertant fractionation.
- r Long-lived **radioactive** isotope. (The decay mode is given if known and the half-life in years.)
- d **Daughter** isotope or decay product of a naturally occurring long-lived radioactive isotope.
- s **Short-lived** radioactive element: only the longest-lived isotope is given. (Half-life in years except when shown otherwise.)

Table B.1
Isotopic compositions of the elements

Element	*Mass number*	*Percentage abundance*	*Note*
H			v,e,m
	1	99.985	
	2	0.015	
He			v,e
	3	< 0.005	
	4	100.00	d
Li			v,e,m
	6	7.5	
	7	92.5	
Be			
	9	100	
B			v,m
	10	19.9	
	11	80.1	
C			v,e
	12	98.90	
	13	1.10	
N			v
	14	99.634	
	15	0.366	
O			v
	16	99.76	
	17	0.04	
	18	0.20	
F			
	19	100	
Ne			v,e,m
	20	90.51	
	21	0.27	
	22	9.22	
Na			
	23	100	
Mg			e
	24	78.99	
	25	10.00	
	26	11.01	
Al			
	27	100	

Table B.1 (*cont.*)

Element	*Mass number*	*Percentage abundance*	*Note*
Si			
	28	92.2	
	29	4.7	
	30	3.1	
P			
	31	100	
Cl			
	35	75.77	
	37	24.3	
Ar			e
	36	0.336	
	38	0.063	
	40	99.600	d
K			
	39	93.258	
	40	0.012	r (β^-, EC, 1.28×10^9)
	41	6.730	
Ca			v,e
	40	96.94	d
	42	0.65	
	43	0.14	
	44	2.09	
	46	0.004	
	48	0.19	
Sc			
	45	100	
Ti			
	46	8.0	
	47	7.3	
	48	73.8	
	49	5.5	
	50	5.4	d
V			e
	50	0.250	r (6×10^{15})
	51	99.750	
Cr			
	50	4.35	d
	52	83.79	
	53	9.50	
	54	2.36	

Table B.1 (*cont.*)

Element	*Mass number*	*Percentage abundance*	*Note*
Mn			
	55	100	
Fe			
	54	5.8	
	56	91.72	
	57	2.2	
	58	0.28	
Co			
	59	100	
Ni			
	58	68.27	
	60	26.10	
	61	1.13	
	62	3.59	
	63	0.91	
Cu			v
	63	69.17	
	65	30.83	
Zn			
	64	48.6	
	66	27.9	
	67	4.1	
	68	18.8	
	70	0.6	
Ga			
	69	60.1	
	71	39.9	
Ge			
	70	20.5	
	72	27.4	
	73	7.8	
	74	36.5	
	76	7.8	
As			
	75	100	
Se			v
	74	0.9	
	76	9.0	
	77	7.6	
	78	23.5	

Table B.1 (*cont.*)

Element	*Mass number*	*Percentage abundance*	*Note*
Se (*cont.*)			
	80	49.6	
	82	9.4	
Br			
	79	50.69	
	81	49.31	
Kr			e,m
	78	0.35	
	80	0.25	
	82	11.6	
	83	11.5	
	84	57.9	
	86	17.3	
Rb			e
	85	72.17	
	87	27.83	r (β^-, 5×10^{11})
Sr			e
	84	0.56	
	86	9.86	
	87	7.00	d
	88	82.58	
Y			
	89	100	
Zr			e
	90	51.45	
	91	11.27	
	92	17.17	
	94	17.33	
	96	2.78	r (?)
Nb			
	93	100	
Mo			e
	92	14.84	
	94	9.25	
	95	15.92	
	96	16.68	
	97	9.55	
	98	24.13	
	100	9.63	

Table B.1 (*cont.*)

Element	*Mass number*	*Percentage abundance*	*Note*
Tc			s
	97	—	(EC, 2.6×10^6)
Ru			e
	96	5.52	
	98	1.88	
	99	12.7	
	100	12.6	
	101	17.0	
	102	31.6	
	104	18.7	
Rh			
	103	100	
Pd			v,e
	102	1.02	
	104	11.14	
	105	22.33	
	106	27.33	
	108	26.46	
	110	11.72	
Ag			e
	107	51.839	
	108	48.161	
Cd			e
	106	1.25	
	108	0.89	
	110	12.49	
	111	12.80	
	112	24.13	
	113	12.22	
	114	28.73	
	116	7.49	
In			e
	113	4.3	
	115	96.7	
Sn			e
	112	1.0	
	114	0.7	
	115	0.4	
	116	14.7	
	117	7.7	

Table B.1 (*cont.*)

Element	*Mass number*	*Percentage abundance*	*Note*
Sn (*cont.*)			
	118	24.3	
	119	8.6	
	120	32.4	
	122	4.6	
	124	5.6	
Sb			
	121	57.3	
	123	42.7	d
Te			e
	120	0.096	
	122	2.60	
	123	0.908	r (EC, 1.2×10^{13})
	124	4.816	
	125	7.14	
	126	18.95	
	128	31.69	
	130	33.80	
I			
	127	100	
Xe			e,m
	124	0.10	
	126	0.09	
	128	1.91	
	129	26.4	
	130	4.1	
	131	21.2	
	132	26.9	
	134	10.4	
	136	8.9	
Cs			
	133	100	
Ba			e
	130	0.106	
	132	0.101	
	134	2.42	
	135	6.59	
	136	7.85	
	137	11.23	
	138	71.70	

Table B.1 (*cont.*)

Element	*Mass number*	*Percentage abundance*	*Note*
La			e
	138	0.09	(r?)
	139	99.91	
Ce			e
	136	0.19	
	138	0.25	
	140	88.48	d
	142	11.08	
Pr			
	141	100	
Nd			e
	142	27.13	
	143	12.18	d
	144	23.80	d,r (α, 5×10^{15})
	145	8.30	d
	146	17.19	
	148	5.76	
	150	5.64	
Pm			s
	145	—	(EC, 17.7)
Sm			e
	144	3.1	
	147	15.0	r (α, 1.06×10^{11})
	148	11.3	d,r (α, 1.2×10^{13})
	149	13.8	r (α, 4×10^{14})
	150	7.4	
	152	26.7	
	154	22.7	
Eu			e
	151	47.8	
	153	52.2	
Gd			e
	152	0.20	r (α, 1.1×10^{14})
	154	2.18	
	155	14.80	
	156	20.47	
	157	15.65	
	158	24.84	
	160	21.86	

Table B.1 (*cont.*)

Element	*Mass number*	*Percentage abundance*	*Note*
Tb			
	159	100	
Dy			e
	156	0.06	
	158	0.10	
	160	2.34	
	161	18.9	
	162	25.5	
	163	24.9	
	164	28.2	
Ho			
	165	100	
Er			e
	162	0.14	
	164	1.61	
	166	33.6	
	167	22.95	
	168	26.8	
	170	14.9	
Tm			
	169	100	
Yb			e
	168	0.13	
	170	3.05	d
	171	14.3	
	172	21.9	
	173	16.1	
	174	31.8	
	176	12.7	
Lu			e
	175	97.40	
	176	2.60	r (β^-, 3×10^{10})
Hf			
	174	0.16	r (α, 2×10^{15})
	176	5.2	d
	177	18.6	
	178	27.1	
	179	13.7	
	180	35.2	

Table B.1 (*cont.*)

Element	*Mass number*	*Percentage abundance*	*Note*
Ta			
	180	0.012	r (?)
	181	99.988	
W			
	180	0.13	
	182	26.3	
	183	14.3	
	184	30.7	
	186	28.6	
Re			
	185	37.40	
	187	62.60	r (β^-, 7×10^{10})
Os			e
	184	0.02	
	186	1.58	d
	187	1.6	d
	188	13.3	d
	189	16.1	
	190	26.4	
	182	41.0	
Ir			
	191	37.3	
	193	62.7	
Pt			
	190	0.01	r (α, 6×10^{11})
	192	0.79	r (α, 10^{15})
	194	32.9	
	195	33.8	
	196	25.2	
	198	7.2	
Au			
	197	100	
Hg			
	196	0.15	
	198	10.1	
	199	17.0	
	200	23.1	
	201	13.2	
	202	26.7	
	204	6.8	

Table B.1 (*cont.*)

Element	*Mass number*	*Percentage abundance*	*Note*
Tl			
	203	29.52	
	204	70.48	
Pb			v,e
	204	1.4	
	206	24.1	d
	207	22.1	d
	208	52.4	d
Bi			
	209	100	
Po			s
	209	—	d (α, 103)
At			s
	206	—	d (α, EC, 32m)
Rn			s
	222	—	d (α, 3.82d)
Fr			s
	223	—	d (β^-, 22m)
Ra			s
	226	—	d (α, 1600)
Ac			s
	227	—	d (β^-, 21.6)
Th			e
	232	100	r (α, 1.41×10^{10})
Pa			s
	231	—	d (α, 3.25×10^4)
U			v,e,m
	234	0.0055	d,r (α, 2.47×10^5)
	235	0.720	r (α, 7.1×10^8)
	238	99.275	r (α, 4.51×10^9)

Bibliography

Anders, E., Hayatsu, R., and Studier, M. H. (1973). Organic compounds in meteorites. *Science* **182**, 781.

Anders, E. and Ebihara, M. (1982). Solar-system abundances of the elements. *Geochimica cosmochimica Acta* **46**, 2363.

Bowen, H. J. M. (1979). *Environmental chemistry of the elements.* Academic Press.

Brimhall, G. H. (1987). Preliminary fractionation patterns of ore metals through Earth history. *Chemical Geology* **64**, 1.

Brown, G. C. and Musset, A. E. (1981). *The inaccessible earth.* George Allen and Unwin.

Bruland, K. W. (1983). Trace elements in the oceans. In *Chemical Oceanography,* vol. 8 (ed. Riley and Chester). Academic Press.

Burbidge, E. M., Burbidge, G. R., Fowler, W. A., and Hoyle, F. (1957). Synthesis of the elements in stars. *Review of Modern Physics* **29**, 547.

Burcham, W. E. (1979). *Elements of nuclear physics.* Longman, London.

Cameron, A. G. W. (1973). Abundances of the elements in the solar system. *Space Science Review* **15**, 121–46.

Clayton, D. D. (1968). *Principles of stellar evolution and nucleosynthesis.* McGraw-Hill.

Clayton, R. N. (1978). Isotopic anomalies in the early solar system. *Annual Review of Nuclear Particle Science* **28**, 510–22.

Cotton, F. A. and Wilkinson, G. (1988). *Inorganic chemistry,* 5th edn. John Wiley and Sons.

Davies, P. C. W. (1986). *The forces of nature,* 2nd edn. Cambridge University Press.

Deer, W. A., Howie, R. A., and Zussman, J. (1966). *An introduction to the rock forming minerals.* Longman.

Duley, W. W. and Williams, D. A. (1984). *Interstellar chemistry.* Academic Press.

Dyson, J. E. and Williams, D. A. (1980). *Physics of the interstellar medium.* Manchester University Press.

Evans, A. M. (1987). *An introduction to ore geology,* 2nd edn. Blackwell Scientific Publications, Oxford.

Faure, G. (1986). *Principles of isotope geology,* 2nd edn. John Wiley and Sons.

Fergusson, J. E. (1982). *Inorganic chemistry and the earth.* Pergamon.

Fowler, W. A. (1984). Experimental and theoretical nuclear astrophysics: the quest for the origin of the elements. *Review of Modern Physics* **56**, 149–79.

Friedlander, G., Kennedy, J. W., Macias, E. S., and Miller, J. M. (1981). *Nuclear and radiochemistry.* John Wiley and Sons.

Ganapathy, R. and Anders, E. (1974). Bulk compositions of the moon and earth, estimated from meteorites. *Proceedings of the fifth Lunar Science Conference* 1181–206.

Goldschmidt, V. M. (1954). *Geochemistry.* Clarendon Press, Oxford.

Greenwood, N. N. and Earnshaw, A. (1984). *Chemistry of the elements.* Pergamon.

Gribbin, J. (1985). *In search of Schrödinger's cat.* Corgi Books.

Gribbin, J. (1987). *In searth of the Big Bang.* Corgi Books.

Grossman, L. and Larimer, J. W. (1974). Early chemical history of the solar system. *Review of Geophysics and Space Science* **12**, 71–101.

Hearnshaw, J. B. (1986). *The analysis of starlight.* Cambridge University Press.

Henderson, P. (1982). *Inorganic geochemistry.* Pergamon.

Huheey, J. (1983). *Inorganic chemistry.* 3rd edn. Harper and Row.

Ihde, A. J. (1964). *The development of modern chemistry.* Harper and Row.

IUPAC (1983*a*). Atomic weights of the elements 1981. *Pure and Applied Chemistry* **55**, 1101.

IUPAC (1983*b*). Isotopic composition of the elements 1981. *Pure and Applied Chemistry* **55**, 1119.

Jacoutz E., *et al.* (1979). *Proceedings of the tenth Lunar and Planet Science Conference, Geochimica cosmochimica Acta* **2**, 2031.

Lefort, M. (1968). *Nuclear chemistry.* Van Nostrand.

Li, Y-H. (1981). Ultimate removal mechanisms of elements from the ocean. *Geochimica Cosmochimica Acta* **45**, 1659–64.

Li, Y-H. (1982). A brief discussion on the mean oceanic residence time of elements. *Geochimica Cosmochimica Acta* **46**, 1671–5.

McElhinney, M. W. (ed.) (1976). *The Earth: its origin, struture and evolution.* Academic Press.

Mason, B. (1971). *Handbook of Elemental Abundances in Meteorites.* Gorgon and Breach, New York.

Mason, B. (1975). The Allende Meteorite—Cosmochemistry's Rosetta Stone? *Accounts of Chemical Research* **8**, 217–24.

Mason, B. and Moore, C. B. (1982). *Principles of geochemistry,* 4th edn. John Wiley and Sons.

Meyer, C. (1985). Ore metals through geologic history. *Science* **227**, 1421.

Murdin, P. (1987). The supernova in the Large Magellanic Cloud. *Contemporary Physics* **28**, 441.

Pearson, J. M. (1986). *Nuclear physics.* Adam Hilger.

Phipps, D. A. (1976). *Metals and metabolism.* Oxford University Press.

Puddephat, R. J. and Monaghan, P. K. (1986). *Periodic table of the elements.* 2nd edn. Oxford University Press.

Reeves, H. (1981). Recent developments in the problem of the origin of the solar system. *Philosophical Transactions of the Royal Society, London* **A303**, 369.

Ringwood, A. E. (1979). *Origin of the Earth and Moon.* Springer Verlag.

Ronov, A. B. and Yaroshevsky, A. A. (1969). Chemical composition of the Earth's crust. *American Geophysics Union Monograph No. 13,* 37–57.

Ross, J. E. and Aller, L. H. (1976). The chemical composition of the sun. *Science* **191**, 1223–9.

Sanderson, R. T. (1967). *Inorganic chemistry.* Reinhold.

Seaborg, G. T. and Loveland, W. (1987). Superheavy elements. *Contemporary Physics* **28**, 33.

Selbin, J. (1973). The origin of the chemical elements. *Journal of Chemical Education* **50**, 306, 380–7.

Shannon, R. D. and Prewitt, C. T. (1969). Effective ionic radii in oxides and fluorides, *Acta Crystallographica B* **25**, 925.

Shannon, R. D. and Prewitt, C. T. (1970). Revised values of effective ionic radii. *Acta Crystallographica B* **26**, 1046.

Smith, J. V. (1977). Possible controls on the bulk composition of the earth. *Proceedings of the eighth Lunar Science Conference* 333–69.

Smith, J. V. (1979). Mineralogy of the planets: a voyage in space and time. *Mineralogy Magazine* **43**, 1–88.

Spinrad, H. (1987). Comets and their composition. *Annual Review of Astronomy and Astrophysics* **25**, 231.

Suess, H. E. (1987). *Chemistry of the solar system.* John Wiley and Sons.

Taylor, R. J. (1970). *The stars: their structure and evolution.* Taylor and Francis.

Taylor, S. R. and McLennan, S. (1985). *The continental crust: its composition and evolution.* Blackwells Scientific Publications.

Trimble, V. (1975). The origin and abundances of the chemical elements. *Reviews of Modern Physics* **47**, 877–976.

Wanke, H. (1981). Constitution of terrestrial planets. *Philosophical Transactions of the Royal Society, London A* **303**, 287.

Wayne, R. (1985). *Chemistry of atmospheres.* Clarendon Press, Oxford.

Weeks, M. E. and Leicester, H. J. (1968). *Discovery of the elements,* 7th edn. Journal of Chemical Education.

Weinberg, S. (1977). *The first three minutes.* Andre Deutsch.

Wilkennig, L. L. (1978). Carbonaceous chondritic material in the solar system. *Naturwiss.* **65**, 73.

Woosley, S. C. and Weaver, T. A. (1986). The physics of supernova explosions. *Annual Review of Astronomy and Astrophysics* **24**, 205.

Element index

The following index gives page numbers for specific elements and their compounds. In addition to these individual references, groups of elements e.g. *alkali metals, noble gases,* or *lanthanides,* are given in the subject index. Also each of the following pages shows a large number of elements:

in tables: 3–5, 130, 183–6, 188–97

in diagrams: 7, 9, 12, 16, 17, 31, 39, 47, 104, 106, 109, 128, 133, 143, 149, 153, 155, 157

Subject index

For reference to individual elements, *see* Element index.